T0064601

MEGAPHYSICS II;
AN EXPLANATION OF NATURE

MEGAPHYSICS II;
AN EXPLANATION OF NATURE

THE EQUATION OF EVERYTHING
IN TERMS OF COSMOLOGY,
STRINGS AND RELATIVITY

DR. MITCHELL WICK

authorHOUSE®

AuthorHouse™
1663 Liberty Drive
Bloomington, IN 47403
www.authorhouse.com
Phone: 1 (800) 839-8640

Published by AuthorHouse 07/23/2015

ISBN: 978-1-5049-1807-7 (sc)
ISBN: 978-1-5049-1808-4 (e)

TABLE OF CONTENTS

FOREWORD AND INTRODUCTION

In this author's first book "Mega physics, A New Look at the Universe", it was postulated that space time curvature followed the spiral fractal formula and was used to locate the quantum ground state in a Relativistic Universe. This is due to the rotational component of the open "flat" expanding universe (Friedmann Type II) and the rotational vector referred to Godel's Rotating Universe.

The area of maximum rotation occurred at Planck's Time 10-43 sec and gradually flattened out with the accelerated expansion after "The Big Bang" until it reached a curvature equaling the space-time curvature metric of Albert Einstein based on the Action Formula S=-1/2k2(-g)1/2R where k is the gravitational coupling constant S is the action of any metric g, and R is the curvature variant on space-time caused by the metric g.

There are two super-symmetric manifolds of space-time as dictated by Quantum Field Theory and brought to light in this author's second book "The Equation of Everything" as -1/2e-in cot theta and -1/2 e in cot theta which correspond to l n (u) and –l n(u) which when superimposed come to flat space time corresponding to one. One component respponds to clockwise unwinding rotation to "The Big

Swirl" and the other to the counter-clockwise rotation of "The Big Swirl" or anti-Swirl. As pointed out in "The Equation of Everything" theta is the angle of trajectory of "The Big Bang" and n is the number of dimensions .e=2.71828 and i=(-1)^1/2.

BACKGROUND

In this author's first book "Mega physics, A New Look at the Universe" the geometry of space-time was postulated as spiral. This is due to the combination of expansion and rotation where if a slice of space-time is made through an expanding and rotating object is cut and the infinite sum is calculated; it is shown that space-time is spiral and follows the spiral fractal formula of Npi/2i $\int_u^{du}$ *where i is the square root of* -1. *This is also equals* $-\frac{1}{2e}$ *– incotangent theta as the integral of* $\frac{du}{u}$ *is* 1n *u. Over the past eight decades, postulates have been* made for the origin of this universe, including "The Big Bang" and "Inflation" theories. Also during this time string theory and M Theory have been devised and evolved as the "New Physics of the 21st Century". Despite this myriad attempts made by physicists have failed to show which type of String Theory consistently explains all phenomena in this universe. This includes the existence of six dimensional curled-up manifolds called Calabi -Yau Manifolds. The hybrid of all five string theories is M theory (Membrane Theory) which look at all five string theories from different vantage points. An analogy is if five blind men are describing a horse from different positions relative to the horse where all five are correct. Despite this all descriptions of the horse are different. As a result all five descriptions are considered dual to each other as are all five string theories. Einstein was able to show that that space-time curvature is a function of mass as a matrix or manifold in which an added surface or surfaces relate to a fixed number of planes that are either moving or stationary in a predictable pattern. Space-time manifolds are displaced by varying degrees of mass and one can visualize a -ball on a trampoline where the trampoline is displaced by the mass of the -ball. Only the trampoline is on every surface of the ball and the ball is curved by everything in the region of the trampoline while the trampoline is curving or distorting everything around the ball. Gravity is the curvature of space-time caused by mass. Gravity is conformal with space-time and the conformations are determined by the mass exerted on space-time exerted through gravitons and fermions (although spin 2 vector bosons are also considered relating to gravity or the curvature of space-time.) Every point in curved space with reference to time as the fourth dimension is attracted to every other point as dictated by the mass of the point with reference to all other points. Also inertia is the resistance to push (or pull)by the mass, so every point resists push or pull as per the inertial (or resting)mass. This can be described by the Ricci Tensor used as a description for resting mass.

Space-time can be considered a metric or g as it acts with a vector and scalar component as conformal to the relationship or inertial mass acting upon it as gravity R g ab for the inertial mass R ab. The definition of a manifold based on surfaces added to a three dimensional surface has, in the past been related or restricted to three or greater dimensions. However, with the advent of flat matter and, elliptical Josephson vorticies, noted as one or two dimensional if dilated time is considered and is congruous to any matter that is superheated, then the definition of a manifold must be changed to include all dimensions of one or greater, the first dimension describes time and the second describes a closed flat string. Einstein predicted that space-time is flat in the absence of mass and the geodesics of a space-time metric g ab has been described as the Lorentzian metric in which it interacts as another metric. When the expression of inertial mass(push)and gravitational curvature of space-time caused by the inertial mass where the mass is described as a metric on the remainder of the space-time manifold are equal, the scalar of space-time manifolds is zero and the manifold is flat(zero curvature).

It has been determined by this author that Space-time is directly proportional to space and inversely proportional to mass. The dimension of perceived time slows down as the object measured approaches a heavy mass. At the event horizon of any black hole space-time shrinks towards zero without actually reaching it as an asymptotic limit. Also space-time is directly proportional to space as in this universe or space-time manifold time's arrow points forward not backwards. Therefore space-time=space/mass x constant and the constant is 1/c2 as space/mc2 where mc2 is the Grand Unification Energy of 10 19 Giga electronvolts is the equation of "The Big Bang" at Planck Time(10-43 seconds). when space-time is -1/2 e –in cotangent theta where n=number of dimensions and the dimensions approach infinity as an asymptotic function ;the expression for space-time approaches zero without reaching it as in the event horizon of a black hole. This is the quantum ground state in a Relativistic Universe and unifies Quantum Mechanics and Relativity.

CHAPTER ONE

BACKGROUND AND STRING THEORY

Strings are the smallest units postulated at Planck Length which is 10-33cm. These strings are basically two dimensional units of either energy or matter depending on which source you read and move in multiple planes. Strings can twist turn, rotate and connect or disconnect with other strings. It is postulated that are made of two types ;open and closed. They move in 26 dimensions which are compactified (rolled up and curled)into either 10 or 11 depending on whether the reader incorporates Supergravity in Quantum Field Theory incorporating gauge symmetry groups as illustrating the 11th dimension. Space-time is described in units called orbifolds; which are manifolds or surfaces which twist and turn in the configuration of a modified cone. A closed string represents a graviton particle or gravitational movements which are mimicked or represented by a spin 2 boson. The world or universe can be represented as a two dimensional sheet(The World Sheet)of either closed or open flat strings that vibrate and rotate with reference to themselves and each other in different combinations which represent twisted or torsed donuts or toruses in flat combinations which can project into multiple planes forming six dimensional twisted or puckered Calabi Yau manifolds. The two dimensional world sheet mapped topologically with conformal mapping must apply the rules of symmetry as well as quantum vibration and have to be dealt with even in a relative vacuum even if this vibration self-annihilated instantly.

Because of the inability to empirically measure string activity, some of the physics community looked askance at string theorists. Despite this, physicists such as Brian Green are trying to quantify string theory with respect to 'The Big Bang' as it was initially broached by Edwin Hubble or 'The Inflation Theory 'as postulated by Alan Guth. Both theories were ex nihilo or "out of nothing" and were postulated as occurring at approximately 13.7 billion years ago although one must differentiate between Hubble Time and Conformal Time as differentiated by Einstein. Both theories were tied in with the expanding universe as discovered by Edwin Hubble in the late 1920's.

String Theory was originally purported in the 1980's when it was discovered mathematically that nature follows harmonics of musical notes. These harmonics were incorporated into two dimensional energy components of matter called strings. Closed strings are continuous and formed loops, double tori, torus configurations(as mentioned previously), triangles, rectangles and a myriad of other configurations based on the energy of the string with regard to other strings in space-time. As mentioned previously the unit of space-time or geodesic is called an orbifold which was defined earlier. Again this unit of space-time is a manifold or surface that twists and turns in myriad configurations as do closed and open strings.

Calabi Yau manifolds are derived configurations of orbifolds and are generally Planck Length or 10-33 cm. Open strings were coined by Dr. Roger Penrose as twisters another description of closed and open strings. Twister theory incorporates imaginary numbers with dimensions to indicate matter in a shadow universe, similar but mathematically different from this's author's -1/2e-incotangent theta incorporated into space-time with regards to quantum mechanics. At Planck Length space-time is curved in and around itself like mini-black holes giving it infinite curvature and transforming it into quantum foam from continuous space-time according to quantum theory while Einstein purported that space-time was continuous down to an infinitely small size. Also, Einstein stated that what was what appeared to be the force of gravity was in fact space-time curvature caused by mass. This is why R g ab is the space-time curvature metric which describes the effect of graviton space-time R causing curvature. String frequencies are based on the note frequency f=1/2L(T/m)1/2 where L is the length of the string(guitar string not string which is smallest unit of matter) (T is the tension and m is the mass of the string. Decrease the length of a guitar string and the frequency increases. Place a finger at the midpoint of the string and the frequency splits in two. These concepts are the basis of string theory's correlations with harmonics or as Pythagorus stated "the music of the spheres" which was first thought of by the Greeks. Open strings split and join and closed strings mimic the spin 2 vector bosons(as previously mentioned)which are quantized units of gravity or mass acting on curving space-time geodesics. The spin 2 vector boson curve space-time as the carrier particles for each and every

3

target mass with the Ricci Tensor R ab. describing the resting mass of the metric g ab. Again this is the effect of gravity rather than it being a force according to Einstein. Hadrons were explained as strongly interacting elementary particles. These spin 2 vector bosons appear as closed loop strings and as a basic building block of matter spin 2 vector boson pervades all mass everywhere and curve space-time as that mass Dimensionality of the space in which strings vibrate relate to1-(D-2)/24 which will only follow relativistic covariance if string moves in a net of 0 dimensional space where a total of 26 non-compactified dimensions exist in Quantum Field Theory and with the SO(32) gauge symmetry group this problem was cleared up partially. C.Everett Peat p.99ClosedStrings appeared in type II string theory like massless bosons with a spin of 2. This relates to the spin2 vector boson involving the effect of gravity. Are strings flat matter or energy? The answer is probably both. One could refer to the Higgs Boson which was coined "The God Particle" which would bear out the boson as being the fundamental building block of matter and energy all pervading. As previously mentioned there are five disparate string theories which have duality to each other. They are type I, type II, type IIa, Heterotic 8x8 and the SO(32) string theories which when observed all as one unit forms Superstrings are groupoids of strings with the mathematical equations that describes the way a sting moves and vibrates, making sure they follow the rules of relativity, quantizing the equations of the relativistic string, and making sure that all strings are supersymmetric and follow the rules of gauge symmetry with strings associated with gauge symmetry groups involving elementary particles. Quantum Field

Theory involves covariant systems where an action can move from one coordinate system to another as in parity involved with the CPT Theorem which involves covariance. The tension of a string was determined as 10^39tons Peat p.101 making it appear to be more energy than matter with a miniscule mass Although the m^2 operator for 2 πT *for the open closed strings determine it has a mass.* $m^2 = 2\pi T \Sigma n = 1 \infty \Sigma = i = 1$ *to* $D - 2a - n^{\mathrm{lanl}}$ *where* $2\pi T$ $= \dfrac{1}{Regge}$ *Slope which is* $a^{1.2\pi T}$ *is huge number but it's sreciprocal is a small number.* Supersymmetry involves properties of iso-spin and strangeness of subatomic particles. M Theory which is short for Matrix Theory or Membrane Theory depending on the source Heterotic strings share two dimensions in one;one rotating clockwise the other counterclockwise with regard to a rotating orbifold or Calabi Yau Manifold. M theory contains supergravity in the 11th dimension(D=11)n the low energy limit and reduces to the type IIa string theory which when compactified forms a sphere. The infinite momentum limit of the D-0 brane may relate to the U(N)super Yang Mills Theory The D-0 brane or membrane relates to the 10 which broke up or cleaved into a dimensional state which has a limit of N approaching infinity for 10 dimensional 0-branes. Based on this argument there were an infinite number of dimensions in the vacuum pre Big Bang state and the 0-branes were the building blocks of everything which were cleaved or broken into a six dimensional component and a four dimensional component where the former is the Ca;abi Yau Manifold and the latter is four dimensional space-time leading to the 10 compactified dimensions of Type IIa string theory. The idea is consistant with the Law of Conservation of Dimensions which states that the sum total number of dimensions of a system in any form is a constant. Infinite momentum(p) relates to the Unified Energy with super gravity in the 11th dimension as described in YangMIlls U(N). To be compactified into a circle or sphere type IIa string theory would indicate world sheet which was two dimensional but spherical or circular in two dimensions. Closed strings would fit better in a world sheet which was compactified to a circle and this would eliminate duality as the five string theories would curl up the world sheet to a point with infinite space-time curvature or a circle of quantum foam when the environment is below Planck Length or 10-33 cm.

CHAPTER TWO

WHAT IS SPACETIME?

Space –time is a term that Albert Einstein coined for what he considered four dimensions of length, width, height, and time. It was based on the Line Element $ds2=dx2+dy2+dz2-c2dt2+dr2$ where r=space-time curvature metric described by the tensor R g ab. Ds2 is a description of the four coordinate system with regard to space-time curvature and the relativistic effect of c2dt2. R g ab or r is determined by R ab or the Ricci Tensor describing inertial mass of an object doing the curving and the curving is done by the spin 2vector bosons and possibly gravitons or fermions. Spiral space-time has a k=-I to the n cotangent theta power as suggested by Sir Roger Penrose and proposed by this author. Flat space-time is space-time without any curvature and occurs in a vacuum state. Considering the fact that photons have a measurable mass while in a moving state moving photons can also curve space-time as energy c, however any mass at Planck Length or greater will curve space-time. It is unknown if mass under Planck Length will curve space-time as there may be a lower limit of Planck Length for space-time to appear in anything instead of quantum foam or quantum dots which may be uniform or curved inward or outward depending on the activity of string components. This quantum foam would make up the orbifold unit of space-time.

Any mass curves space-time from string sized to that supermassive black hole which centers each and everyone of the 750 billion galaxies in this universe and can be displaced by the spiral or cone shaped designation of space-time as it approaches the event horizon of a black hole where perceived time appears to the intelligent observer to shrink along with local space to a point which may also be string sized or 10-33 cm at the center of the black hole although there is no proof that the center of a black hole isn't greater in size than that of a string.

Einstein's equation of Relativistic Gravity described the Einstein Tensor G ab=R ab-1/2R g ab=8(pi)T ab where the Einstein Tensor=Ricci Tensor-1/2(Relativistic Gravity)=8(pi)T where T=stress energy tensor of the metric g ab. Relativistic Gravity is the curvature of space-time caused by the metric g ab and the Ricci Tensor represents the inertial mass of the metric g ab. Inertia-gravity =1/2 but with the metric tensors being abelian and anti-symmetric Inertia-gravity and gravity-inertia are equal but opposite in magnitude and direction netting out zero with is the value of the Einstein Tensor revealing a stress energy tensor of approximately zero.

The equation was derived by the Lorenzian Transformatios but basically says stress energy is zero in the "Pre-Big Bang" or "Post Big Crunch "epoch. In black holes the gravitational effect from a collapsed neutron star or galaxy is so great even light can't escape.

Beyond the event horizon of a black hole space-time is collapsed in a spiral or vortex configuration to almost zero(as mentioned previously) and mass is collapsed to almost infinite density. This mimicks the pre-Big Bang where the Einstein Tensor G ab approaches zero. This is illustrated in the tensor expression of The Equation of Everything where R abcd=R abc-1/2 R g ab/R ab where space-time is curved inward by the collapsed mass of a black hole. In essence, the space-time curvature metric is gravity acting on R, which is the spa ce-time that is curved by the metric g ab. In the "Pre-Big Bang" epoch(up to 10-43 seconds or Planck Time)the space-time curvature metric approaches infinite curvature as a point which has infinite curvature and the extreme mass of the pre-Big Bang quantum bubble curves extremely small space-time toward infinity without reaching it. In this case gravity approaches half of a very large non-infinite number and anti-gravity from antimatter approaches a very large non-infinite number but inertia equals the sum total of the gravitational and antigravitational metric acting on space-time with almost infinite curvature. This all indicates that Gravity isn't a force but the effect of space-time curvature caused by any mass. The action of any metric in s[ace-time R with the curvature metric of (-g)1/2. R can be described in terms of dimensions such as d to the nth dimensional power and is extremely useful in string theory with its postulated 26 dimensions compactified(curled up)to 10. As previously mentioned Relativistic Gravity is described by 8(pi)Tab=R ab-1/2R g ab where R g ab describes Relativistic Gravity and its effect on space-time. The space-time curvature metric is described in the tensor equation R a b c d=R a b c -1/2r g a b/R a b where R a b c d describes curved Lorenzian space-time R a b c describes flat Mintkowski or Riemannian space and R g ab describes the space-time curvature metric acting upon R a b c causing the curvature caused by the mass described in the Ricci Tensor R ab describing the inertial mass of metric g ab. The entire expression is multiplied by 1/c2 to give the Relativistic form of the Equation of Everything. Utilizing Einstein's equation of Relativistic Gravity and this author's equation of everything and solving the stress energy tensor T ab one gets the gravitational constant G6.67x10-11 newtonmeters/sec2. Setting R ab from both equations equal to each other by the transitivity postulate a=b b=c therefore a=c Tij==T ab.T ba where i=initial event and j=final event and mass=energy/c2 as T i j=

T ab. Tba=Tij/||c2||2 and 8 pi T=R ab-1/2 R g ab via Einstein's formula where 8(pi)T ab. T ba/c4 reflects the stress energy tensor T ij=R ab-1/2R g ab at zero stress energy. In other words the Einstein Tensor G ab=T ab. T ba=G(the gravitational constant)as 8(pi)G/c4 is the gravitational coupling constant k and the stress energy of zero has constraints of the pre-Big Bang or post-Big Crunch and at event horizons of black holes. Stress energy is approached by the gravitational constant G=6.67x10-11 newton meters/sec2 which approaches zero at the point of the Pre Big Bang" epoch or where the Einstein Tensor approaches zero. The difference is attributed to weak perturbations as described by the action formula S or the Hamiltonian Operator for n eigenstates of energy as the 0 or null eigenstate is approached. An operator is a complex function which operates on another function such as the LaPlacean Operator, which acts as a second degree differential equation of the function it is operating onto the upward limit of the operator. An eigen-state is a state of matter or energy with regard to the Operator that operates on another function such as the quantum level of matter with regard to energy.

CHAPTER THREE

THE BIG BANG; WHAT WAS IT?

OVER 13.7 BILLION YEARS AGO a quantum bubble with a 50:50 mix of matter and antimatter underwent either a 360 degree orb blast, inflation, or a two phase swirl which continuously expanded with progressive decrease in rotation. Prior to that is up to speculation. There is one theory that two membranes from other universes collided or touched triggering "The Big Bang' or a series of zero-BRANES the building blocks of matter underwent a dimensional recombination like "the popped bedsheet" hypothesis of Dr. Michio Kaku 1 forming a six dimensional manifold in a Calabi Yau configuration and the macroscopic four dimensional manifold of space-time. Another hypothesis is that a previous universe underwent a "Big Crunch" causing time's arrow to reverse and turning positive time into negative time(backwards)due to the implosion until it reached just before time zero when time advanced to Planck Time 10-43 seconds and the matter-antimatter mix exploded. The gravitational effect of antiparticles and particles may have caused "The Big Bang" and resulted in Dark Energy which propelled the accelerated expansion of galaxies away from each other that Edwin Hubble discovered in the late 1920's. According to the math antiparticles repel each other causing such an explosive force in just under 50 percent of the matter antimatter mix that from a Planck Length string sized quantum bubble the "Big Bang" results in a massive antigravity surge from antiparticle antiparticle repulsion. This phenomenon not only explains the existence of Dark Energy with pushes galaxies apart from each other but also the paucity and lack of cohesion of antiparticles in the universe.

Antimatter was first discovered in 1932 and since then myriad antiparticles have been discovered. Antiparticles are being isolated in Cern, Switzerland in the Hadron Collider but as of this date the mass of any. antiparticle has not been conclusively discovered. It has been noted that antiparticles such as the positron have the opposite charge as matter particles but it has not been determined that antiparticles have positive gravity. There are some experiments that do show repulsion of antiparticles in positive gravity fields 2, but to collide particles and antiparticles can cause annihilation of the particles with the expulsion of energy but this doesn't necessarily mean that particles and antiparticles attract each other by gravity;only possibly by charge and electromagnetism which are stronger forces than gravity which acts as a weak force but isn't actually a force as previously mentioned. As antiparticles self repel the energy required to push antiparticles toward each other would be considerable and might destroy the antiparticles. If antiparticles mutually repel with antigravity instead of attract with gravity a plausible mechanism for "The Big Bang" can be made from the quantum bubble. At Planck Time 10-43 seconds a 50:50 mix of matter and antimatter caused an orb blast with a trajectory of theta in the spacetime equation -1/2e to the i n cotangent theta power where i=the square root of -1. And no is the number of dimensions either 4 for macroscopic spacetime or 10 compactified dimensions including Calabi Yau Manifolds or Oribfolds in string theory. The 2 pi radian orb blast with the mutually repulsive force of antiparticles forcing an explosion converting over 99.999 per cent of the antimatter into Dark Energy by the formula E=mc2 postulated by Albert Einstein where m=mass of the antimatter. Whether the mass is positive or negative is up to speculation but the Law of Conservation of Energy was purportedly been violated by the "Big Bang" jnless the potential energy of the quantum bubble equaled the kinetic energy, heat, and Dark Energy after "The Big Bang". However the potential energy of the quantum bubble didn't spontaneously appear "ex nihilo" or out of nothing and may have been from a collision of a matter universe, antimatter universe(or membrane) and a relative vacuum forcing the pre-Big Bang implosion which drove time's arrow backwards and reducing entropy from two universes to a quantum bubble(making the occurrence a singularity as the Second Law of Thermodynamics and Time's Arrow pointing forward must be suspended for this act) As antiparticles repel dark energy pushed outward in all directions carrying the balance of matter with which it is subsequently attracted to other matter by gravity but not to antimatter(which appears to have a lack of cohesion);the formula R j I k l-Rjl kl(where kl is a superscript to the R in the second term while in the first R j I k l are subheadings all to indicate covariant and contravariant tensors respectively=-8(pi) {Ge ji}}= g ji in four dimensions of space-time(an opposite curvature or reciprocal curvature to gravity being antigravity as the sign for—(8 pi{[Ge j i]}=g j iis negative. The rightside of the equation is the mutually repulsive force of antiparticles and g j I is the metric of the antiparticle where i=initial event and j=final event and-8(pi)G where G=the gravitational constant is from the Cosmologic Constant ^which reflects the mutually repulsive force pushing galaxies apart purported due to Dark Energy.E=2.71828 and e j I is the vector product of e from j to i. This assumes a 360 degree or 2 Pi radian orb blast in the "Big Bang" and R is the anti-gravitational effect on particles while conversely R j i is the antigravitational effect on antiparticles. In a 360 degree or two pi radian orb blast the angle of trajectory is pi radians or 180 degrees. The cosine of pi radians =-1 which explains the -8(pi)G on the right side of the equation. There are approaching an infinite number of 180 degree slices in a perfect sphere so the angle of trajectory for an isotropic universe must be 180 degrees or pi radians.

THE C.P.T. THEOREM AND ITS' INNATE SYMMETRY OF NATURE

Charge, parity and time whether reversed or not have innate symmetry in nonlocal systems according to Quantum Field Theory. This indicates that if charge were reversed as in the positron vs. the electron the magnitude of the charge would be essentially unchanged. If time were reversed that $\Im\Psi(x,t)\Im-1=eiT\Psi(x,-t)$ *where* $i=\sqrt{-1}$ *and e is to the iΦ power, the transformation of t to –t can be shown as communicative as the operator &&& is antiunitary. The operator cancommute with the Hamiltonian.*

The Hamiltonian and still reverse the sign of t.3 With parity X can be replaced with –X and still have a commutative Hamiltonian Operator such that (CPT) $\mathcal{H}(CPT)-1=\mathcal{H}(-x)$ *-time curvature of matter and antimatter. Combining matter and antimatter have space -time curvatures which would complement each other cancelling each other out resulting in flat space.*

Space-time when matter and antimatter annihilate each other. This causes energy=mass of the antiparticle+mass of particleXc2 resulting in the interfitting of the reciprocal curvatures of space-time for identical particles and antiparticles resulting in asymptotic flatness. In an antimatter universe of manifold pre-domeinantly anti-matter anti-gravity would attract rather than repel repel anti-particles due to Parity replacing X by –X with the commutative Hamiltonian Operator and in this case gravity would repel rather than attract particles for the same reason. Conversely in a matter dominated universe anti-matter would repel anti-particles with anti-gravity and matter would attract particles with gravity. Outside of weak perturbations the symmetry of nonlocal systems is upheld with the C.P.T. Theorem.

CHAPTER FOUR

TIME'S ARROW AND THE
LAW OF ENTROPY

Time's Arrow states that time will be move forward in our space=time continuum. The Law of Entropy or the Second Law of Thermodynamics states that every system or subsystem will always go from a more ordered state to a less ordered state. This obviously occurs in an open flat expanding universe, but what happens in the event that there is a "Big Crunch"? In an implosion where a less ordered state goes toward a quantum bubble where space-time develops a rip or tear space=time can "pop" like a balloon and the second law of thermodynamics may be violated. Mathematically, the equation space-time=space/mass(1/c2) negative space-time equaling negative space/positive mass x 1/c2 which dictates the rate of compaction of space-time in a "Big Crunch". Also according to Einstein's Law of Relativistic Gravity G a b=R a b-1/2R g ab where G a b=0 in a pre-Big Bang or post Big Crunch epoch. Here R a b stays positive for the Ricci Tensor or inertial mass but the space-time curvature metric known as gravity changes sign or the direction of the vectors of the positive mass from –R g ab to +R g ab and the stress energy tensor 8(pi)T a b goes equal in magnitude but opposite in direction. Due to the fact that space-time=space-time and with the Bianchi Identity -1/2R g a b=1/2 R g ab in magnitude but opposite in direction but since anti-symmetric they net out to zero so the inertial mass of this universe stays the same as the Ricci Tensor R ab. Since the only way to prove these conclusions is in a Big Crunch which could occur in an accelerating rate toward Planck Time 10-43 or slowly where the only proof would be a shift toward the ultra-violet with regard to the decelerating expansion of galaxies as measured on the Hubble telescope. As approaching a heavy mass time slows down the mass/per space decreases as in a Black Hole time must slow down in a Big Crunch toward a quantum bubble and in that case "Time's Arrow" would decrease. The equation $\mathbb{R}$ a b c d=R a b c -1/2R g ab/R a b changes to $\mathbb{R}$ a b c d=R a b c+1/2R g ab/R a b in the event of a Big Crunch as space-time undergoes reciprocal curvature from the space-time curvature metric called gravity in other word space-time would curve inward in the presence of inertial mass instead of outward and with space-time reducing in size to a point or quantum bubble such as is analogous to a black hole event horizon R a b c+1/2 R g a b is greater than R a b c. The entire expression R a b c+1/2 R g a b is greater than R a b c d which is decreasing and as R g ab=-R g ab as anti symmetric tensors then R a b c d=-1(R a b c-1/2R g ab/R a b) whuch indicates the decreasing direction of time's arrow with the curved Lorenzian or Riemannian Space-time $\mathbb{R}$ a b c d.

BLACK HOLE ENTROPY IS BASED ON STEVEN HAWKING'S FORMULA S=2(pi)(NQ1Q5)1/2 where s= entropy N is the number of states or eigenstates in a black hole postulated as 252 separate states and Q1and Q5 are the differential charge between the first and fifth eigenstates. S=degree of disorder. The one brane in M theory(membrane theory)is described with the monopole fixed negative charge suggestive of an electron or positron with the antiparticle and the five-brane represents 4 –space or curved Lorenzian Space-time which spiral into a black hole event horizon. As a result the energy of a monopole acting on curved Lorenzian Space-time across 252 states of matter reveal black hole entropy. Also S(black hole)=Area/4 Length2 P=c3A/4Gh where h=Planck's Constant P=Planck Length of 10-33 mA=cross sectional area and G is the Gravitational Constant of 6.67 x10-11 newton meters/sec2 as c=3x10 8 meters /second or the speed of light which is a true boundary for any mass as space-time shrinks to approach zero at that boundary. Dr. Hawking postulated the entropy of a black hole to be 0.29 which approaches zero. This indicate that as the cross sectional area approaches 0 or P(Planck Length)the entropy of a black hole approaches zero(0)according to the Bekenstein-Hawking Equation. This mimics the entropy(S) of the Quantum Bubble at pre-Planck Time before "The Big Bang" where a 50:50%0 mix of matter and antimatter are solidified by enormous pressure into what might be called a lattice formation much as a diamond would occur. Indeed the central locus of a black hole post event horizon ight have the same or similar configuration as with tremendous pressures strange matter and liquid states of matter which would under other ambient conditions not be liquid or solid. In a "Big Crunch" which may occur in Planck's Time(10-43 sec)which is fast or a slow leak of space-time like a deflating balloon would eventually approach 252 eigenstates of a quantum bubble similar to that of a black hole and measured time would slow down as a shift toward the ultraviolet would occur as galaxies recede instead of expand. A nuclear clock might lose 10-3 seconds for each month that the galaxies recede instead of expanding and would be for most in perceptible except with scientific measurement. Also there would be no clear indication as to when the tail end of a "Big Crunch" would occur in the last 10-43 seconds where everything would shrink to approximately 10-33 cm like the initial quantum bubble. As a result it is extremely difficult to empirically prove that" Time's Arrow" is reversed in a "Big Crunch" also because the energy to reverse the sequencing of events in a universe would require the energy of a Big Crunch.

An alternative explanation for the end of this universe would be "Heat Death" where all matter and antimatter would slow down its' expansion until it eventually stops all fusion and fission in stars would eventually slow down and stop and everything would slow down to a crawl at 2.74 degrees kelvin(the temperature of the background microwave radiation from the "Big Bang")and all matter would "freeze".

CHAPTER FIVE

WHAT IS SCHWARZCHILD SPACE-TIME AND HOW DOES IT RELATE TO BLACK HOLES?

It is well known that as the event horizon of a black hole is approached space-time approaches zero, time is dilated toward infinity(infinitely long)and mass increases dramatically along with gravity as space-time approaches a point string sized or 10-33cm in de Sitter or anti-de-sitter space. Quasars develop and spume out matter energy(Hawking Radiation)4 and information from the event horizon of a black hole whose origins or poles depend on the location, velocity and mass of the observer and are based on observational viewpoints.

The entropy of a black hole was already discussed in the previous chapter and again was postulated by Hawking as 0.29 where at least 252 different states of matter appear as N in the equation S=2(pi) {NQ1Q5)1/2 where Q1 is the membrane boundary of the monopole(as mentioned previously)representing an electron cloud bounded by the membrane known as the 1-BRANE in terms of M Theory and Q5 represents Lorenzian Curved space-time or 4 space on which the 1-brane binds it with the electron(or positron)cloud acting upon it across the 252 states of matter. Q1 is 1.602x10-19 coulombs or the charge of an electron and a unit of space-time referred again as the oribifold and is generally bounded by Planck Length(10-33)cm. Therefore black hole entropy=2(pi)(252x10-19coulombs)(10-33 cm) to the one half power or 2(pi)(4.032x10-54 to the one half power or 2(pi)(4.032x10-108)

which approaches zero entropy. At the point at the event horizon the conformation is spherical and two dimensional with regard to position and action of the observer and the observer's motion. The mass of the information at the Event Horizon relates to its' spherical radius(which decreases toward zero)in Region II of collapsing matter using Schwarzchild Space-time. The event horizon is basically a cork which is a spherical region with extremely dense mass froma collapsed neutron star or galaxy and the hole of the event horizon is basically clogged up by the mass. In a way that matter queues up in that spherical region of collapsing space-time that approaches zero or string sized. Region 1 of Schwarzchild Space-time has a constant radius and constant time. Region 2 has a radius approaching twice the mass and time approaching infinite dilation. Region 3 has the radius approaching twice the mass with time infinitely dilated in the negative direction. In other words time's arrow is reversed as in the tachyon. Region 4 has increasing time with regard to past and future time cones. The isotropic coordinates of Schwarzchild Space-time Metric is from the line element ds2=-(1-M/2r)2 divided by 1+m,/2r)2 and dt2+(1+m/2r)4{dr2+r2)d(omega)2 where omega is the Hubble Constant. Using the Kuskel Extension of Schwarzchild Space-time one has a symmetrical hourglass or cone shaped(spiral) configuration with the throat of the hyper surface at t=0 or infinitely dilated with the radius=twice the mass which is actually a 2-sphere or two dimensional hypersurface with one dimension suppressed. The topologic configuration of the hypersurface is RxS2 and is a circle shown with the radius equaling twice the mass. The surface above the throat at radius=2M lies in Region 1 and below the throat with r=2M in Region 4. There is a hyperbolic region proximal to the throat at the event horizon where r approaches 0 and time approaches infinite dilation. It has previously been determined by this author and the behavior of tachyons that negative mass goes backwards in time and reverses time's' arrow. Region 3 must be composed of negative mass as time's arrow is reversed as time approaches infinite dilationto the negative side while with ordinary matter(positive mass) it approaches time to positive infinity. SUBSTITIUTING a –MASS for positive mass in region 3 and reflecting it back to region 2 of Schwarzchild Space-time the t=infinity and t=-infinity cancel to time=0(zero)and the r=2mand r=-2m(minus 2 Mass)at zero space-time at time=0(zero), it can be concluded that mass and its energy equivalent cancel in regions 2 and 3 and only regions 1 and 4 are left. The information is not lost or destroyed but cancelled out with the negative mass cancelling the positive mass at the event horizon. The quasar effect is scattered throughout space from the 2 dimensional hypersurface over 2(pi)radians or 360 degrees of arc from a point in spacetime of zero entropy and dilated time over string sized space-time into curved Lorenzian Space-time with entropy being increased according to the second Law of Thermodynamics. The above is an explanation for "The Hawking Paradox" which states that the information absorbed by a black hole is lost(energy and mass)when the black hole eventually evaporates.

The negative mass must also produce antigravity and reciprocal curvature of space-time in region 3 which when superimposed on region 2 will push region 3 into region 2 in a form equal but opposite to region 2 resulting in asymptotic flattnessin regions 1 and 4. If one substitutes –mass for +mass in ds2(1-m/2r)2divided by (1+m/2r)2dt2 gets ds2=(1-m/2r)2divided by (1+m/2r)2times (1+m/2r)2divided by (1-m/2r)2 squared dt2. Taking the square root of both sides with negative mass substituting for positive mass with the Schwarzchild metric one gets ds2=1(dt)2or ds2=dt2 as the dx2+dy2=dz2 of the line element cancel out with the negative and positive masses at the event horizon with the negative and positive exteme masses at the event horizon from region 2 and region 3 being reflected upon region 2. This is consistent with space-time approaching zero at the event horizon and stops time at a point of Planck Length where the positive mass from region2 and negative mass of region 3 meet. Note in the cone approaching region 2 and region 3 from the opposite direction the radius gradually approaches zero and the mass and negative mass are sequestered proximal to the event horizon so the extreme progressive curvature of space-time to the infinite curvature of a string sized point is preserved. Therefore the net information going in and out of a black hole event horizon is a symmetric bi conar surface with region 3 cancelling region 2 leaving regions 1 and 4 to produce the quasar effect.(See diagram 1b in appendix)

CHAPTER SIX

WHAT IS THE 'EQUATION OF EVERYTHING'?

There exists a simple mathematical relationship between space time and mass relating to gravity(space-time curvature metric) and the speed of light. In terms of a verbal description the relationship is simple. In terms of Relativity and Quantum Mechnics, the relationship is more complex.

As an object approaches an area of extreme mass such as a black hole, time slows down and eventually stops. As any object with mass approaches any other mass with is larger, time slows down even infanitesimilly. As a space-time with an atomic clock would approach the star(Solaris)or the sun, an internal clock would lose at least $1/10^{th}$ of a second, possibly more. And if possible to approach a black hole, time would slow down through dilation towards zero where time would stop. Also space-time=space-time therefore space-time=space-space-time curvature metric(known as gravity) with inertial mass expressing the curvature. Therefore space-time is directly proportional to space. Also space-time is inversely proportional to mass as is proven by the action of space-time as it approaches the event horizon of a black hole. Therefore, space-time is directly proportional to space and inversely proportional to mass. The actual equation would be space-time=space/mass times a constant. This constant is $1/c^2$ or 1/the speed of light squared as energy=mc^2 and the denominator having mc^2 becomes the Grand Unification Energy at the point of the "Big Bang" incorporating everything.

In terms of tensors R a b c d=R a b c-1/2 R gab/R a b where R a b c d is curved space-time or Lorenzian Space-time R a b c is flat Mintkowski or Riemannian Space-time R g ab is the space-time curvature metric known as gravity for the metric g ab whose inertial mass is described by R a b which is the Ricci Tensor of that mass. That ENTIRE EXPRESSION IS MULTIPLIED BY THE CONSTANT $1/c^2$ to give the equation of everything.

The Equation of Everything in terms of Relativity as postulated by Albert Einstein is that all motion is relative and not absolute. When mass(m)travels at approaching the speed of light boundary inertial mass approaches infinity, space-time approaches zero(0) and length shortens to infinitely short or perhaps Planck Length according to the Lorenzian Transformations. As item A of mass m travels west in an environment that is traveling at velocity B when item A is traveling at velocity A the total velocity is the sum of A plus B. with vectors equaling the components of the motion away from 2 pi radians or 180 degrees such as (A plus B)cosine theta where theta is the angle in radians which is the difference between 2(pi)radians and the net angle displacement of A and B with regard to the surface or manifold. If there is another manifold or surface which is traveling at velocity C which if positive is added to velocities A and B cosine theta If C is negative or traveling less than velocity(magnitude and direction)A and B then velocity C is subtracted from velocities A and B cosine theta. If the manifold is moving in an expansion with a trajectory of 180 degrees as in the "post Big Bang" cos 180 degrees is one so the result would be velocity A and velocity B cos theta +or – the velocity of space-time with regard to the stationary observer. The Equation of Everything is spacetime=space/mass times $1/c^2$ or $\mathbb{R}abcd$ = Rabc-1/2R g ab/R ab all times $1/c^2$. In terms of the metric g ab Lorenzian Curved Space-time or Riemann space-time is $\mathbb{R}$ a b c d as previously mentioned. R a b c describes flat space-time on which the metric of gravity R g ab curves space into curved space-time(as previously mentioned). The metric of gravity emanated from mass m whose inertia is described by R ab(Ricci Tensor)and this curves flat Mintkowski Space-time either inward or outward depending on the mass being acted upon by the metric of the mass doing the curving. The sum is described as R g ab where g a b is the metric of mass m.

The apace-time curvature metric of Einstein emanates from Einstein's Equation of Relativistic Gravity where a progressively increasing mass as described by the Ricci Tensor R a b-1/2 the gravity or space-time curvature metric which also increases with increasing mass as space-time curvature approaches infinity as in a point as in the space-time of the pre Big Bang quantum bubble(if below Planck Length)then a quantum foam described by zero-Branes (as previously mentioned). This depends on whether de Sitter Space has a hard boundary at Planck Length as in String Theory's Oribifold. The orbifold again is a twisted cone which is complex and can twist into a CalabiYau Manifold or surface which is a continuous surfacein a puckered appearance of a double torus that communicates with other Calabi Yau Manifolds 5 all six dimensional and in motion.

As increasing inertia and increasing gravity do not increase at the same rate as the speed of light boundary is approached space-time progressively curves to a point at v=c with infinite curvature. According to the Lorenzian Transformations infinite mass in dilated timeand reducing space-time with progressive increase in curvature causes inertia to increase at a greater rate than gravity because the metric g ab is acting on progressively increasing space-time curvature which is R g ab and the space-time curvature metric is half the inertia because the tensors are anti symmetric, abelian and space-time follows Bianchi's Identity. The covariant and contra-variant tensors of space-time are abelian with regard to the space-time curvature metric and as mentioned before are anti-symmetric as inertia approaches infinity at velocity approaches "c" without reaching it which is when the Einstein Tensor G ab=0 which is the stress energy tensor T ab at infinite space-time curvature. Anti-symmetric tensors cancel out in magnitude but with opposite direction.

THE EQUATION OF EVERYTHING IN TERMS OF QUANTUM MECHANICS HAS space-time=space/massx1/c2 described in terms of Planck Mass(the smallest unit mass for a quantum particle)or that two quanta can occupy and is approximately 1.22x10-24 kg. Mathematically Planck mass is the square root of hc/8(pi)G where h=Planck's constant at 6.63x10-34 and G is the Gravitational Constant of 6.67x10-11 newton-meters/sec2 and c=3x10 8 meters/sec and is the speed of light boundary. Space-time is described as the n-Dimensional state of a point-particle x at time t as a probability function and is operated on by the Hamiltonian Operator defined as –h/2mtimes the La Placean Operator with respect to the second derivative or d2/dx2+d/dy2+d2/dz2/ So utilizing the above space-time of H a}(r,t)|2 this is the probability density of a point particle "r" with respect to time or "t" in n-Dimensional Space. This equals -1/2e to the + or – I to the n cotangent theta power, here i=the square root of -1or -1e is the inverse or reciprocal of the natural log which is the integral of du/u which relates to the spiral fractal formula introduced in this authors first book "Mega physics, A New Look at the Universe" and this defines the ground state in a Relativistic Universe as the natural log (l n 1)=0 and the natural log of infinity=infinity. Theta is the angle of trajectory at Planck Time from "The Big Bang" which is pi radians or 180 degrees. The infinity power of e(2.71828)is infinity and e to the negative infinity power is zero. Therefore e –in cotangent theta power defines the ground state as ln 1=0 and e I n cot theta is a reciprocal function and describes 1/0 which is infinity but e –in cot theta is zero where n=number of dimensions and theta is the angle of trajectory or pi radians. Therefore Planck's Mass(c2)H a|(r, t)|2 d n r where this expression is multiplied by the La Placean Operator in the n-dimensional state. A is the number of eigenstates of energy operated on by the Hamiltonian Operator (-h 2/2m)x La Placean Operator. In this case the Operator defines the wave function of r with respect to time(t) and also explains weak perturbations which explain quantum fluctuations in deSitter Space. Note also that the Hamiltonian Operator-minus h2/2mtimes the LaPlacean Operator)2+V o or initial velocity reflects momentum p=mv where p(rho) is Momenetum. So the momentum of quanta with Planck's Mass is incorporated over n-diemsnions from a to n eigenstates of energy and weak perturbations must include the momenta of quanta from the a=0 to a=n eigenstates of energy or energy levels. Therefore c2(hc/8 pi G)1/2+H a(eigenstates)|(r,t|)2 d n r times the LaPlacean Operator represented by the inverted delta to the n power equals =-1/2e +or –i n cotangent theta power with fluctuations which is again described by the Hamiltonian Operator in "a" eigenstates in the n dimensional state and ei n cot theta power is infinity. Therefore as 10 19 GEV approaches infinity. here n is the number of dimensions and I is the square root of -1. The Hamilton Operator in n dimensions describes the wave function of a point particle r with respect to time(t) in the n dimensional state with respect to "a" eigenstates of energy.c2(hc/8 pi G)1/2=10 19 power Giga Electron Volts which is the Grand Unification Energy which includes all forces except weak perturbations from quantum fluctuations with are corrected for by the Hamiltonian Operator. This occurs in the multiverse where the n-dimensional state approaches infinity and can be proven by subdividing a sphere to one second of arc or 1/3600 of a degree. This second of arc can be subdivided down to infinitiy and each subdivided portion is in motion as part of space-time with an infinite number of intersections of each unit or infinitely subdivided second of arc. As the intersection of two planes define a dimension and since the subdivided portions are not parallel due to space-time curvature when any even an infinitesimal amount of mass(non-vacuum)curves all space-time there are an infinite number of intersections from these subdivided lines or planes indicating an infinite number of dimensions in the multiverse.

When one second of arc is a plane in motion the topological surfaces mimic an osculating plane 7and each osculating plane from a topological standpoint define a dimension as space-time curvature causes an infinite number of intersections of these infinite osculating planes. According to Zeno's Paradox each degree is subdivided ad infinitum that prevent two solid objects from touching. With an infinite number of dimensions for space-time the left side of the quantum mechanics equation narrows from 10 19 giga electron volts toward infinity without ever reaching it as is true with the right side of the equation. There are also other theories such as M Theory which states the zero-branes which were building blocks to everything may have incorporated an infinite or near infinite number of dimensions. This situation applies for space-time to -1/2e –in cot theta and this applies to everything with it's reciprocal -1/2e i n ot cot theta power and this applies to zero space-time on the right for -1/2 e –I n cot theta. On the left side the Planck Mass is zero in the ground state as this is the vacuum or null state so c2(hc/G)1/2=0 and the zero-dimensional state states that the Hamiltonian Operator or point particle|(r,t)|2 d n r times the La Placean Operator in the zero dimensional state is also zero because the derivative of zero is zero. Therefore 0=0 in the ground state and the Equation of the Universe or Multiverse is upheld with respect to Quantum Mechanics. Note that this expression equals the Relativity Expression ℝ a b c d=R a b c-1/2R g a b/R a b times 1/c2 and due to the transitivity postulate that a=b,b=c therefore a=c indicates that the Quantum Mechanics Equation which is zero in the ground state and the Relativity statement which is zero in the ground state(as R a b c d=0 as Riemannian space nets zero when mass or inertial mass R a b approaches zero. Of course R a b c-1/2R g ab=0 ; R a b approaches zero forms the expression 0/0 which is everything as the case of curved Lorenzian Space-time. This is the case of where approaching the 0 dimensional case incorporates everything.

Kurt Godel described a scientific tenet called "The Axiom of Incompleteness" stating that in any axiomatic system the set containing all elements must be incomplete. If there exists a set containing this subset, the axiomatic system must be incomplete. Based on this tenet "An Equation of Everything" must be incomplete although it appears complete. An example of this is when 10 19 Gigaelectron volts(The Grand Unification Energy of all energies from "The Big Bang")can only approximate the value of infinity and while it is true the 10 19 Gev approximates infinity due to weak perturbations from quantum fluctuations it isn't definitive. However in the infinite dimensional case where 0=0 as 10 28 power in the denominator of the left side and infinity in the denominator of the right side approximate 0=0. This will again be brought into focus later. Quantum Mechanics and Relativity are unified with the equations ℝ a b c d=R a b c-1/2R g ab/R ab times 1/c2 and Planck Mass times the expectation value of the probability of a point particle r at time t in the n dimensional state where space-time is -1/2 e –i n cot theta with the reciprocal curvature being -1/2e i n cotangent theta where n=the number of dimensions netting infinity divided by infinity which is everything except zero which is the null state or absolute vacuum state indicating in this total case that nothing or spaceless ness doesn't exist. Based on measurements the integral of du/u mathematically indicates -1/2 e –i n cotangent theta power as space-time with the curvature metric R g ab while the sum total of both reciprocals for space-time would result in flat space-time as in a vacuum which is precluded by the expression infinity/infinity because it doesn't include zero and is therefore incomplete. The Quantum Mechanics equation of everything is $c2\left(\frac{\hbar c}{8\pi G}\right)1$ $\frac{\psi(r,t)dn(power)r\nabla n}{}$ *H from* $+$ *a to n eigenstates (|r, t) | d n(power)r* ∇ $=\Psi(r,t)d$ *(n power)r* ∇ *(n power)* $-\frac{1}{2}e-$ *in cot* θ equals the Relativity Equation R a b c d=R a b c-1/2R g a b/R a b times 1/c2 and at the ground state they both equal zero and each other and can be simplified to space-time=space/ mass all times k=1/c2 where c=3x10 8 meters/sec.

The Hamiltonian operator $\hbar\frac{2}{2m}\nabla$ *to the nth power where* ∇ *(nabla) is the LaPlacean Operator handle the weak perturbations and*

or quantum fluctuations acted upon by the wave function of the point particle r at time t. This will be delved into again later in this book.

Another equation postulated was the wave function $\psi(r,t)=\int e^{\,i}_{\,h}$ to the integral power $\int(\frac{R}{16\pi G}+\frac{1}{4F2}+\psi iD\psi-\lambda\varphi\psi\psi+D|\varphi|2-V(\varphi)$ *where* Ψ *is bar* Ψ *or a probality function. This is based on the Yukama Equation, Relativity,*

33

The Schrodinger Equation and the book "Quantum Mechanics and Path Integrals by Richard P. Feynman and Albert R. Hibbs. The path integral $\oint \nabla 2\psi\left(r,t\right)$ *applies Poisson's Equation for the path of dual vector field $4\pi\rho$ where ρ is the energy*

Energy density of matter and *$\psi(r,t)$ is the wave function of point particle r at time tforming a path integralof $16\pi\rho2\psi(r,t)$ or 16π*

$16\pi\{\frac{\rho3}{3}) - \frac{}{3\ (r,t)\,superimposed}$ *with parallel transport to curved manifold (surface)with space-time curvature of* $-\frac{1}{2}e-$ *in for cotangentθ for* $\frac{\rho3}{3}$ *and space-time curvature of* $-\frac{1}{2}e+$ *in cotangent theta for* $-\frac{\rho3}{3}$ *again approaching the ground state. The-$|D\phi|2-V(\phi)$relates to the Higgs Field which relates to the spin 2 vector boson which curves space-time by any mass as described by the Ricci Tensor.*

CHAPTER SEVEN

WHAT IS M THEORY?

M Theory is a combination of the five extant string theories ;type I, Type II, Type IIa, Heterotic 8x8 and the SO(32) string theories. The five string theories are dual to each other as previously mentioned mathematically observing the same phenomenon from five different vantage points or approaches all describing the same thing. What makes M Theory the combined five string theories is the incorporation of membranes which vibrate. These membranes are submicroscopic and may be string sized 10-33 cm or possibly smaller. A membrane is almost continuous with any and all matter and possibly energy and are continuous across the different dimensions whether 26 compactified(curled up)to 10 as in type I string theory or an approaching infinite number of dimensions if the osculating plane can be applied to an infinite number of non-parallel planes which are subdivided from 1 second of arc(1/3600 degree)and in motion expansion and rotation as in Godel's Rotating Universe. These non-parallel planes must intersect an infinite number of times if they are non-parallel due to the space-time curvature metric forming an infinite number of dimensions in the osculating planes. Membranes would have to be contiguous with these osculating dimensional planes and would contain all matter including energy which is converted to matter as matter=energy/c2. M Theory was originally coined for Membrane Theory and Matrix Theory depending on the source read and the different membranes called in short hand branes describe different states of matter interacting with energy in space-time. M Theory has no clear cut definition except the duality with all five string theories although when utilizing super gravity for the existence of the 11th dimension instead of the compactified 10 dimensions in the low energy limit involving D-0-Branes it reduces to type II a string theory when compactified by $2\pi R$ *where R measures the limit to infinity of D-0-Branes in the infinite momentum limit of MTheory as in the case of U(N)in the super Yang Mills Theory*

Theory 9. The compactified type IIa string theory is a sphere of Radius R. The D-0-Brane has a momentum limit of 1/R and a bound state of D-0_branes have a conditional momentum of N/R where N relates to the unified Yang Mills state relating to the gauge limit in terms of d4space such that g2YM N approaches infinity. Note that D-1-Branes or 1-branes describe the string as well as the monopole depending on the text read. M theory incorporates super gravity and the 11th dimension to string theory. "branes" are movable membranes that incorporate special dimensions and particles including strings. Membranes as mentioned previously vibrate move and are real and measurable. Charges, state and tension are incorporated into membranes which extend up to at least the ten dimensions of string theory. The limited definition of M Theory is "the limit of strongly coupled IIA string theory with 11 diemsnional supergravity or the Poincare invariance" The 2-brane or M 2-brane couples to the potential(V) of eleven dimensional supergravity. The 5-brane or M-5 brane carries an electromagnetic charge of the potenetial(V)of eleven dimensional supergravity. The 5-brane carries the potential which is coupled to the 2-brane. The Neveu-Schwarz 5-brane carries the electromagnetic charge(V)of what is known as the NS-NS 2-form potenetials where the NS boundary state is a fermionic field that is anti-periodic on the world sheet in the closed string or the double of the open string. The Neven-Schwarz algebra is a world sheet algebra with regard to the energy momentum and super-current tensor into a sector where supercurrent is antiperiodic and modes are half integer valued. Therse are what are known as Fourier modes. The null state is orthogonal to all physical states including its' own. Incorporating the anti-periodic fermionic field which is the NS boundary state onto the 4 and 5-brane representing space-time in the four macroscopic dimensions would have the Neven-Schwarz 5 brane incorporated with the NS-NS 2-form potentials on the world sheet (which represents flat or two dimensional matter) and is divided by the Ricci Tensor with regard to the sum total or infinite sum of all inertial mass to equal the Lorenzian tensor of space-time in terms of supergravity incorporating 11 dimensions. The momentum limit of the inertial mass as it approaches the infinite momentum limit where "R" approaches a very large number is incorporated into the denominator of space-time=space/mass. Space-time curvature based on the infinite momentum limit R where N approaches infinity in the unified Yang Mills state in the d4 state relates to the macroscopic gauge limit and the D-0-brane which is bound with conditional momentum of N/R. The momentum limit of 1/R of the D-0-branes relates to the infinite momentum limit and curved Lorenzian space-time or Riemannian space-time where the 256 permutations net out to zero in the ground state. Applying this to 11 or possibly more dimensions(infinite dimension state as previously mentioned in terms of M theory is more difficult.

SUPERGRAVITY which incorporates the 11th dimension into superstring or the IIa closed string theory to compose M Theory where the IIa closed string theory compactifies to a circle or sphere is involving global supersymmetry which produces a Yang Mills Gauge Group of Osp(1/4). The number of states within 11 dimensional supergravity are eAM implies ½(9)x(10)-1=44

$$\psi\, Mimplies \frac{1}{2(9x32-32)} = 128 \quad and \quad AMNP \begin{pmatrix} 9 \\ 3 \end{pmatrix} = 84.$$

where M and N represent 11 dimensional curved space indicies with 32 dimensional spinors. The term e A M where A is the superscript and M is the subscript and it represents a linking or parallel transport of the base manifold or surface with the tangent space.

ψM is the graviton field and A MNP is an antisymmetric tensor field with 128 *boson fields equalling the fermin* equalling the fermionic field giving a total number of 256 boson and fermionic fields which is the number of fields in the 11 dimensional N=8 model for super gravity again where bosons and fermions curve space-time.

As there are 256 bosonic and fermionic fields curving flat Mintkowski or Riemannian space-time which has 256 permutations and since

Riemannian space-time graviton or bosonic fields are anti-symmetric tensor fields acting on flat space-time as anti symmetric tensor fields the net permutations of gravity space-time curvature on flat space net out to zero which is the numerator of space/mass=space-time. Regardless of the denominator even when multiplied by the speed of light squared the net result is zero which is curved Lorenzian or Riemannian Space-time. Space-time in the "null state" or equivalent to the Einstein Tensor G ab. Therefore "the Equation of Everything" does apply to 11 dimensions as incorporating supergravity. Note in the above Riemann Space-time can be applied as curved or flat when Mintkowski Space-time is always flat. Riemann postulated that all mass can be described as curves in space-time although flat space-time means that a missal shot out in flat space will theoretically never return to the starting point but with curved space-time it will eventually return to the starting point as in Einstein's Closed Curved Universe vs. Friedmann type II open flat expanding universe.

In terms of M-Theory the Neven-Schwarz 5- brane incorporated with the NS-NS 2 potentials incorporating the anti-periodic fermionic field(vacuum state of ferminon field) on the world sheet would go into the numerator of space/mass and the infinite momentum limit in the bound state of D-0-branes would go into the denominator suggesting the inertial mass times the speed of light squared of the quantum bubble with approximately 750 billion strings acting on D-0-branes in the pre Planck Time epoch of under 10-43 seconds. The world sheet would be 2 dimensional as are strings and the anti-periodic fermionic field density in terms of a tensor field and applying

Poisson's Equation for dual vector fields would be applied via parallel transport onto the flat manifold of the world sheet. Since the energy momentum and supercurrent tensor are also anti-periodic the energy momentum R goes into the denominator with the supercurrent incorporating into the 1-brane as a carrier such as the monopole or electron cloud while the fermionic field density is incorporated in the numerator as previously mentioned. This incorporates to enclose a near infinite momentum onto an accelerating expanding space-time acted upon by gravity (fermionic fields caused by the inertial mass of the quantum bubble). AS A RESULT N-BRANES SUGGESTIVE OF LORENZIAN SPACE-TIME=D-0-BRANES+ and –Neven -Schwarz 5- Brane incorporated with the NS-NS 2 potentials/N/R where R is the infinite momentum limit and N=1. The N-Brane as an infinite number of dimensions are approached but never reached approaches infinity but is never reached and since anything/1/ ∞ *is infinity due to the infinite momentum limit so as applied to N dimensions or the N-brane infinity=infinity with supergravity localizing the Einstein Tensor G ab which is* $8\pi T$ *which is the stress energy tensor*

Energy tensor of matter acting by the Poisson Equation and is zero(0) based on the D-0-branes.

CHAPTER EIGHT

WHY THERE IS AN ARGUMENT THAT THERE ARE AN INFINITE NUMBER OF DIMENSIONS

A dimension is formed by the intersection of two planes which have an intersection of an angle which in length, width and height is 90 degrees or pi/2 radians. However other dimensions may have different angles of intersection which aren't perpendicular. For example, the intersection of the x,y, and z axis has a near infinite number of points which while discontinuous can be subdivided down to an infinite number of intersections and the limit of these subdivisions can blend or blur into a continuous dimension. There are a postulated 26 dimensions associated with string theory and superstring theory and these dimensions can be compactified(rolled or curled up)to 10 dimensions or 11 dimensions with supergravity included. To make a corollary in logic one must depend on axioms as assumptions. These axioms must be assumed to be true and if false the corollary can be false. For example if it's an axiom that planes are stationary and later it is found that planes are in motion, then the corollary that there are three dimensions length, width and height with no others as in

Euclidian Geometry is false. In the case of space-time this corollary is false as time is considered a fourth dimension by Albert Einstein and his space-time curvature theory which was proven with the "solar eclipse" experiment in 1921 to measure a change in position of an object near the sun(which is of significant mass compared to empty space)proves that space-time is curved and not flat(without curvature). It was mentioned previously by this author that each subdivision of a second of arc(or $1/3600^{th}$ of a degree)can be subdivided down toward an infinite number of cuts. If these cuts of space-time were totally parallel to each other as if space-time were totally flat and without curvature, then these cuts would never intersect each other. However because space-time has been postulated as being in motion along with mass, expanding and rotating(according to this author and Kurt Godel), then these subdivided planes would not be absolutely parallel but would be almost parallel, where the parallel nature is distorted as space approaches a reads of extreme mass such as a black hole where the curvature of space-time increases towards infinity which is why space-time appears to spiral into a black hole in a cone like configuration as the event horizon is approached. The time component or fourth dimensions shrinks toward zero as the extreme mass is approached as space-time is inversely proportional to mass.

Now if there are a near infinite number of planes from the subdivisions of each second of arc and is these planes are osculating and in motion expanding with a progressive decrease in rotation from the area of maximum rotation at just before Planck Time 10-43 seconds and approaches pure expansion with only a miniscule rotation as 13.7

billion years later ;then these planes are osculating into expansion and rotation vectors with the unit tangent vectors being measured along these osculating planes. Obviously, if all these axioms are true, these near infinite(if there is no downward boundary to space-time at below Planck Length or 10-33 cm)and if the quantum foam can also be subdivided down to infinity then there would be an almost infinite number of intersections between the near parallel planes formed by each subdivision from 1 second of arc downward toward infinity. As Einstein showed that these planes of space-time are curved and "The Big Bang" Theory shows them in motion, these are osculating planes and the unit tangent vectors reflecting the curvature of these planes and the planes will intersect a near infinite number of times forming a near infinite number of dimensions. This was all explained earlier in this book and there are other arguments to this premise as well.

This conclusion would be totally consistent with the infinite or near infinite universe theory and would preclude that D-0-branes are not at absolute zero dimensions but an infinitely small number approaching the zero dimensional state as it is when an object approaches 0 degrees kelvin but cannot be reached. Indeed the null state or D-0-branes are only an approximation where the zero dimensional state is approached as an asymptotic function as with limits where the size of each extant dimension is reduced toward zero without reaching zero. The reason why the axiom nothing doesn't exist when according to "The Big Bang" ex-nihilo states it does is because of these limits in size downward for the dimensions which are in motion with the osculating planes never reaching the zero dimensional state as when "time's arrow" reverses time from before the Big Bang when a Big Crunch would cause time to move backwards in a massive implosion bringing it back toward zero time without ever reaching it before the Big Bang moves time's arrow forward again. These subdivisions can be curled up or compactified toward zero BUT ARE NOT ZERO ONLY APPROACHING ZERO AS AN APPROXIMATION. Also the diverse motions of string and superstrings in space-time are diverse but the number of motions from a vibration to a twist, partial rotation(degrees can also be subdivided plus the motions of strings can be subdivided)make the two dimensional world sheet as an approximation as would the two dimensionality of flat matter or flat anti-matter. Indeed, the other dimensions are there but so miniscle and compactified that for the sake of math they aren't important and there limits can be brought to zero.

Infinite Dimension Symmetry involves the symmetry transformation of space-time as. generated by similarity transformations of T ab which is the stress energy tensor of theories associated with conformal gravity C.A type of algebra exists called infinite dimension subalgebra which is associated with Lie Groups as a subgroup of Lie Algebra. Infinite dimension subalgebra has nonsingular commutators and uses gauge symmetry of the Yang Mills groups to form a weighted tensor algebra. An approaching infinite set of conformal fields have well definied commutators of an arbitrary pair of zero modes constituting an infinite dimension subalgebra with regard to the full symmetry of string theory and M Theory(including the compactified IIa string theory which is spherical. Infinite dimensions do not arise through the moding of a finite number of conformal fields but rather than an infinite number of conformal fields of space-time acted upon by any mass. W ∞[17] *in terms of field equations with Wi*[18] *as conformal weight W∞ must retain all modes of the conformal fields of space-time space-time while zero modes form only Cartesisan Subalgebra. This subalgebra possesses the properties of string symmetries*

Ties of string symmetries and is a supersymmtery in that it's generators do not commute with the generators of the Lorenzian Transformations and comingled or quantum groups which are excited and with a different spin it is spontaneously broken in flat space-time because NOT ALL GENERATORS COMMUTE WITH THE STRESS ENERGY TENSOR(T ab)relating to the free scalar and it transforms excitation of differing masses into one another. Symmteries or gauge symmetries should be local symmetries and the constant tensors *ψandχ in the generator were depending on the scalar field Xμ(σ)where mu is a superscript the*

Locality would be apparent, not it is x dependence that affected commutations. Infinite dimension subalgebra would be a global part of Gauge Symmetry and must include both holomorphic and anti-holomorphic derivatives with propagating degrees of freedom generated by operators evenly balanced between two types of derivatives. Short distanc e singularities e ipxx L(x)ei qXL(w)equals the fraction e i(P+Q).xL(W) /(Z-w) to the-p .q power and e ip.x(=L(x) is also a power with e iqXl(w) as a power. The stress energy relates to (Z-w)to the –p.q acting upom e i(P+Q).Xl(W) which relates to e –in cotangent theta of space-time, The infinite dimension state where

Napproches infinity forces the N-brane of M theory to approaches an *∞Brane for Curved Lorenzian Space-time R a b c d and the superequation*

{Ha to n eigenstates|(x,t)2d nr ∇n *divided by*

$$c2\sqrt{} \hbar \frac{c}{8\pi G} = \nabla n |x,t| 2dnx\nabla x\left(-\frac{1}{2e} - in\ cotan\ theta\right)$$ *where n dimensions approaches ∞ therefore* $\frac{1}{2}e - i$ i n cotθ *approaches e-∞power so the*

$$wave\ \frac{functiony|x,t|2d\ \frac{n(powerof)x\nabla n}{-1}}{2}e - in\ cot\ thea = \psi|x,t|2dn\frac{x\nabla 2}{\infty} = 0.$$

Here space-time is described as $-\frac{1}{2}e -$

i n cotθ *where theta approaches π radians or 180 degree trajectoryof a sphere describing* the Big Bang. Note the particular math of infinite dimension su-algebra gets extremely involved and can be referred to by the footnote 10.

Call{ } the set of all covariant tensors in d –dimensional space. The operator *{ψ (k), wi} such that Σl = 1{ψ(h), wi +δil} where ψ is the wave function of w i and here i is the initial eventthe element of algebra of pairs is such that*

Pairs such that{ψ(h),w i}modulate relations→λ{ψh,wi}+ω{ψ(h),w i}-{λψ(h)+ωψ(h),w i→)such that Δh{ψ(h),w i}→0.ψ is a subset of S. In this case h is not Planck's Comstant but a variable power to the wave function. S,η is the Mintkowski Metric for flat space-time.|S| are the elements of S;S-ψ are the complement of ψ M.S.Conformal weights ω5-ψ+V(t)-P(u) were conformal weights of ψ that don't correspond to the indicies in ψ together with the weights of χ

Weights of χ *that don't correspond to the indicies in* ψ *with the conformal weight of* χ *that don't correspond to the indicies in the*

To indicies in the image of the wave function of w and P(U) such that ωs-μ+V T=$P(U)$*Wick's Theorem states the sum over* μ *and P is a sum of all possible contractions of the tensors* ψ

Tensors ψ *and* χ *The conformal weights are those of the noncontracted indicies except the weight*

Of ψ *and are increased incrementally by the weights of the contracted indicies through the power of the operator*

Of the operator $\Delta|S$-$\mu|$*Thet wo point function for free bosons implying the curvature of Mintkowski space-time by any mass is*<*x* $m(z)xV(w)$≥-log *(Z-w)*$\delta\nu\tau o\mu$ *where* δ *is the superscript and* μ *is the subscript and the-*log *(z-w) is to this* δ *where* μ*is the covariant tensor and* δ *is the contravariant tensor all aplied to the* Boltzman Equation incorporating states of matter with regard to bosons and conformal gravity. Wick's Theorem states that short distance singularities as at the event horizon of a black hole form a conformal block or non-holomorphic operator or sum of the products of holomorphic and anti-holomorphic fields constructing a full commutator out of the commutators for anti-holomorphic components. Holomorphic and anti-holomorphic operators are constructed from mutually commuting sets of the creation and annihilation operators. Short distance singularities at Black Hole Event Horizon are eip.xL(z) where ip and x L(Z) are powers times e iqX L(w)=the quotient of e to the i(p+q). xL(w) power/(Z-w)-p.q .p.q must be integers. Products of holomorphic and anti-holomorphic fields constructing a full commutator out of commutators for anti-holomorphic components form the conformal block on space -time|S| giving a mechanism in terms of math on how space-time is crammed down by annihilation operator in terms of bosonic effect of extreme gravity on asymptotically flat Mintkowski Space-time s, *η as event horizon is a black is reached.* This may appear far a field from infinite dimensions but the stress energy tensors relating to the annihilation and creation operators acting on Mintkowski Space-time clearly reveal that space-time or the n-brane =space(D-0-Branes + or – the Neven Schwarz 5-brane with NS-NS2 potentials/N?R where R is the Infinite Momentum potential as N approaches 1(which is mass or the Ricci tensor x c2). They also explain why space-time spirals down to a point (string sized?) without actually disappearing. The Hamilton-Jacobi Equation for infinite dimensions is such that $\lambda v(x) + <Ax = \phi(x), Dv(x)> + H(ABpower\ x, D\ v(x)) = 0$ *where x ε X wher e X is a real Hilbert Space,λ>0 and H:H by X is continuous as a closed linear operation with a compact and dense inclusion* D9A0⊏

D(A) is less than X. We assume A is positive and self adjoint. *$\phi:D(A$ β power) implies D(A-beta power) and is Lipschitz continuous. Piermarco Cannarsa and Maria Elisabetta Tessitore worked on the Type equation here.*

MEGAPHYSICS II CONTINUED;
CHAPTER EIGHT

The infinite dimension Hamilton-Jacobi Equation and applied specific situations is in a paper written by Piermarco Cannarsa and Maria Elisabetta Tessitore called "Infinite Dimensional Hamilton-Jacobi Equations and Dirichlet Boundary Control Problems of Parabolic Type".

The stress energy tensor of a free scalar T a

$$b \longrightarrow T(\sigma') \, and \, T(\sigma') = \frac{1}{2} : \partial x \partial x (\sigma) := \lim \varepsilon 0 \longrightarrow \frac{i}{2 \partial x (\sigma) \partial x (\sigma + \delta)} + \frac{1}{4\pi \delta 2}$$

where the commutator is $4[T(\sigma), T(\sigma'] = \lim \varepsilon,$

$\varepsilon' \longrightarrow 0[\partial x(\sigma \partial x(\sigma + \varepsilon)] \partial x(\sigma') \partial x(\sigma' + \varepsilon')\}$. *for real numbers.or*

F(σδ(σ-σ^')-f^' (σ^')δ(σ-σ^')→[T(σ),T(σ^')] which applies to the commutation of the stress energy tensor where σ and σ^' havethe stress energy

E stress energy as related to T a b where g a b relates to sigma and sigma prime. Incorporating imaginary numbers 2iT(σ')∂'(σ-σ')- iT'(σ')δ(σ-σ^'-ε')/(ε+ε')2+2δ(σ-σ')-δ(σ+ε-σ')/(ε+ε')3. Stress Energy between sigma and sigma prime yields epsilon relating to a small number for stress energy relating to black hole entropy(s) using Virasuro Algebra: e ip .X(σ)::e iq.X(σ+ε)power=:e ip dot X(σ)+iq power dot x(σ+ε):e p-q/2π power. As ε *is greater than zero the entropy of a black hole approaches zero by the normalization coefficient with normal ordered parameters* normalization coefficient with normal ordered operators. The stress energy between sigma and sigma prime is a maximum at the event horizon or a black hole where space-time approaches epsilon.11

As is well known s=r(theta) θ *where s is the distance or arc length subtended by a sphere r is the radius of the cut of the sphere and*

If the sphere is compactified and converted to two dimensions with regard to the world sheet as in type IIA string theory converts to a description in which M Theory may be derived. As theta θ *approaches zero degree but not reaching it the cuts becomeinfinitely small. The radius remains c* remains constant. As the sphere describes space-time in the Big Bang rotating and expanding according the H the Hubble Expansion Coefficient the cuts "s" approach zero initially then expand as theta approaches zero then expands and "r" describes the radius of this universe. If the near infinite number of cuts remained parallel theta would remain very small (below one second of arc). The radii would approach being parallel without reaching it as the angle between radii approach zero degrees without reaching it. The cuts are not parallel due to the extreme mass of the quantum bubble with extreme Dark Energy overcoming the extreme mass of the bubble curving space-time to infinity as in a point which is a two dimensional sphere composed of strings. The unwinding curvature of space-time will reach the space-time curvature metric of Einstein instead of that of asymptotic flat space. These infinitely small cuts with ds approaching zero as theta approaches zero with r or the radius continuously increasing according to the Hubble Expansion Factor forms an infinite number of dimensions as the curvature causes an infinite number of intersections of these non-parallel infinite number of planes formed by the infinite number of cuts in the circle (2D) or sphere (3D) or space-time (4D) and these planes are osculating with R and R. moving along the Normalization component such that arc length parameter r=r(s) with $s = \left\| \int_a^t \left\| \frac{dr}{du} \right\| \right.$ or $\frac{ds}{dt} = \|r\|$ with a dot over r.r.is differentiation with respect to time $r^{\wedge}$ differentiationwith respect to distance ampping t to s has inverse distance.r.=r'=dr/dt,r..=r''=d2r/dt,r...=r'''=d3r/dt. Mapping t to s has inverse relation of s to t given by t=psi prime(s) dt/ds =psi prime(s)=1/‖r dot‖The moving frame is such that T(unit principle normal)r'=(dx/ds,dy/ds,dz/ds)The Binormal vector with a curve B=TxN for the cross product shows all plane curves have a principal normal and these curves are not paralle.N=(-sinθ,cosθ,0)plane z=0 if T=(cosθ,sinθ,0)*The Moving frameat any point r.isn 'tzero but may be episoln and r.r:aren 'tzerp*

zero then T=r(dot)/‖r(dot)‖and N=ε(r.r.)(r..)-(r.r..)r./‖$r^{\wedge}$' ‖‖r.xr..|$\dashv$|

where x is the cross product

This moving frame in the form of a triad moves continuous along C where C is analogous to r is the formula.: s=r θ *of a sphere divided into an infinite number of moving cuts or planes and each C and N are mutually orthogonal*

Are mutually orthogonal the triplet of unit vectors T,N and B constitute a right handed system of basis elements E3 and cover space-time curvature as -1/2e-in cot theta. The triad of T,N and B moves continuously along C and is the moving frame or triad whereby T and N are the osculating plane.(touching unit tangent vectors)12 Now considering compactification of a near infinite number of dimensions or cuts in a circle or sphere, the area can form a point with infinite curvature moving at the rate of the Hubble Expansion factor with the force or energy of the Big Bang. Therefore the infinite intersections of planes in motion outward with a curvature metric going from infinity as a point to space-time curvature metric from any mass as derived from Einstein these dimensions which are the intersection of two moving planes along the vectors T, N and B will compactify below Planck Length or 10-33 cm which is the size of a string leaving at least 10 or possibly 11 non-compactified dimensions as in supergravity or string theory rather than the 26 non-compactified dimensions. The creation and annihilation operators which are holo- morphic and anti-holomorphic can be used to show space-time crunching to an infinite curvature point in a "Big Crunch" from an expanding and rotating sphere and expanded and rotating from a point of infinite curvature to this universe with all infinite dimensions compressed in the infinite curvature point to where space approaches nothing without reaching it and the infinite dimensions all MUTUALLY INTERSECT AT THAT POINT. Again the compactified form of II a string theory becomes a two dimensional sphere or circle which corresponds to that infinite curvature point or expands with two dimensional components such as closed strings to THE WORLD SHEET.

A complex idea being made simple is based on Ockham's (or Occum's) Razor. All things being equal, the simplest explanation tens to be the right one. If a two dimensional sphere is a circle which is the COMPACTIFIED FORM OF II a String theory then the equation *s=r θ where s and θ are subdivided down to infinitely small slices but with a radius that approaches*

Infinite length has a compactified infinite number of dimensions of N-branes approaches the D-infinity Brane where all of the dimensions are compactified in the quantum bubble the dimensionality approaches the D-0-Brane just as *2π radians approaches* 0 *radians on a circle but never reaches it. The radius if approaching infinite*

Infinite length going across 360 degrees or 2(pi) radians could radiates flat matter (2 dimensional matter or radiation)at v approaches the speed of light boundary where space-time shrinks because the inertial mass of matter approaches infinity. A radiating energy source over 2 pi radians from the quantum bubble as the radius before the "Big Bang" would travel at v approaches c. Length=Length (0)(1-v2/ c2)1/2 from the Lorenzian Transformations for matter. There is length contraction at v=c just as there is time dilation where time slows down towards infinity. There is a theory that length was the first dimension de-compactified and would have to be infinite ®R from which the Hubble expansion coefficient would follow but according to the Lorenz Transformations radiation would have to be infinitely short if traveling at v=3x10^8 meters/sec which can happen in collapsed space-time where time is dilated towards infinity and space is compactified to the size of a quantum bubble or 10-33cm(String Sized). Based on this the Grand Unification Energy GUT would explode at v=c while space-time exceeds it to accommodate the mass and energy. Many questions can be answered by the arc length equation *s=r θ and this is a very simple relationship which can be utilized for the compactification* of type II a string theory to aid in the explanation of the BIGBANG.

If arc length or S and theta represent space-time and R or the radius can incorporate everything else it is possible to show space-time=space/ massxc2 relates to s=r(theta). The N-branes where N approaches infinity=D-0-Branes+ or −Neven Schwarz 5-brane-1/2NS-NS 2 potentials associated with Fermionic Field intensity caused by the infinite momentum limit 1/R where the Fermionic Field is gravity or the curvature of space-time caused by R ab which is the Ricci Tensor associated with the metric g ab where the infinite momemtum limit is 1/R or infinity. As the D-0-branes=0 and the gravity space-time curvature metric is infinity when dealing with the infinite curvature of a point where the ferminonic field and bosonic field have 128 permutations each acting on 256 permutations of Riemannian or Mintkowski space-time you get 0/1 at the infinite momentum limit where R approaches infinity you get 0/0 as 1Based on this is ther1/ ∞ *is zero therefore* $\frac{0}{0}$ *is mathematically everything including zero.*

If the radius of s=r θ *is* $\frac{0}{0}$ *or everything including zero and s and θ are space-time there is an argument that this equation could be used for the infinite dimension equation* and to mathematically explain "The Big Bang" another way if the compactified form of IIa string theory is a circle.

CHAPTER NINE

HOW WILL THIS UNIVERSE END?

There is a theory that this universe will end with "Heat Death" whereby the accelerated expansion with progressively decreasing rotation of space-time with the approximately 750 billion galaxies will slow down and eventually stop with the distances such that any gravitational effect caused by galaxies will become negligible and space-time will approach asymptotic flatness except for the effects of the Cosmologic Constant ^ suggestive of reciprocal curvature of space-time caused by Dark Energy almost cancels space-time curvature caused by the sum total of all the mass of each galaxy. When this happens eventually stars will cool down and collapse into black holes where the majority of gravity will be sequestered and the ambient temperature will approach 2.74 degrees kelvin resulting in widespread freezing.

An alternative to this is the "popped balloon scenario" of space-time which may possibly trigger a massive implosion and a "Big Crunch". The speed 3×10^8 meters/sec or approximately light speed "c" is a boundary in which ordinary matter can't seem to breach. Tachyons as subatomic particles seem to be able to breach this boundary as well as going backwards in time(time's arrow reversed). This occurrence can occur if tachyons are bosons with a negative instead of positive mass which would make it impossible to tachyons to travel below the speed of light at which point they would be bosons traveling forward in time and with a positive mass. It has been postulated by this author and Steven Hawking that time's arrow can reverse in a massive implosion or "Big Crunch" or by this author that negative mass as may occur at the event horizon of black holes causing a superimposition of regions in Schwarzchild Space-time netting a solution to the Hawking Paradox as mentioned in a previous chapter. Normally in a Big Crunch which could have a slow phase followed by a pop inward of approximately Planck Time 10-43 there may just initially be a light Doppler shift toward the ultraviolet instead of the infrared showing the accelerated expansion of galaxies away from each other slowing. Note this can also happen in the early state of "Heat Death" therefore measurements would have to show a gravitational metric or curvature of space-time more than canceling the anti-gravitational effect of Dark Energy and the cosmologic constant to warrant consideration about a "Big Crunch" unless the "popped balloon scenario occurs in which case the implosion could be so fast that it would equal but be opposite the explosion of the "Big Bang" with gravity instead of anti-gravity predominating.

Space-time curves inwardly during the singularity of a "Big Crunch" as it would at the event horizon of a black hole. If a "Big Crunch" would occur the end of this space-time manifold could occur by a rip or tear in space-time as the expansion at a progressively increasing rate and rotation at a progressively decreasing rate since "The Big Bang". As previously mentioned the quantum bubble with infinite curvature space-time progressively uncoils towards asymptotic flatness. According to Kurt Godel's rotating universe which is shown by the clumpiness of the WOMP showing the BMR or baseline microwave radiation from "The Big Bang" the clumpiness shows increased uptake and diameter which indicates strong perturbations from the rotational vector of the expansion of the universe. Note that when you take an osculating plane that's rotating and expanding simultaneously and you take a cut of this plane the result is a spiral configuration which is what Albert Einstein postulated as the configuration of space-time in 1913. Note that in an increasing expansion with decreasing rotation space-time can achieve a rip or tear like an inflating pair of pants with a tear that becomes increasingly larger until it pops like a balloon which is a logical conclusion to Cosmic Inflation which is Dr. Alan Guth's 12 hypothesis for the expansion of space-time with conformal gravity. This tear occurs at "c" which is the boundary of the speed of light where dilated time acts as if it stops. The Lorenzian Transformation shows time dilated towards infinity or becomes infinitely long as "c" is approached closer and closer. This lack of space-time at the speed of light barrier reflects the small rip or tear in space-time with inertia being caused by the near infinite mass of the matter approaching the speed of light boundary and the resistance of space-time above light speed causes the rip which will in time increase in dimension until a "Big Crunch" s induced. During a "Big Crunch" slow phase time will slow down then stop and reverse. This only occurs if space-time travels at faster than light and created this boundary such that matter is pushing against a brick wall formed by space-time forming the near infinite inertia shown by the Lorenzian Transformation. It was also postulated that "the speed of gravity" is greater than "c" or the speed of light, but as gravity is the curvature of space-mass caused by mass, gravity reflects the rate of curvature of space-time caused by the mass of the matter approaching the speed of light. As inertial mass for any matter approaching light speed approaches infinity the curvature of space-time at the speed of light boundary approaches infinite curvature or a point such as the tip of a cone at v=c where the cause is the pressure of space-time above v=c. If space-time exists at v is greater than c but shrinks to an infinite curvature point at v=c and then exists with the space-time curvature metric of Einstein at v <c; the speed of light barrier is a weakness of space-time that is similar to a small tear which can increase by the expansion of space-time.

It is much less likely that a "Big Crunch" will occur rather than "Heat Death" because the amount of mass versus space in this universe is very small except at the speed of light boundary or at the event horizons of black holes. If the number of black holes were to increase geometrically due to the implosion of many galaxies without those black holes "evaporating" the the amount of mass would increase compared to pinched off space-time at every event horizon. As it is there many be tears in the inflating balloon of space-time at each black hole event horizon. Despite this the space(3 space) vs mass(matter and antimatter) is so skewed toward empty 3 space (as compared with 4 space or space-time)that the curvature of space-time caused by the extant mass of all 750 billion galaxies +sum of all black holes wouldn't cause sufficient curvature of space-time to cause significant rips or tears; and since the ratio of 3 space to mass would increase as long as the acceleration of galaxies continue over time or in space-time the likelihood of a "Big Crunch" would decrease vs. Heat Death. If space were rife with black holes (active or dry)and if space-time was continuously collapsing in multiple spirals then the gravitational curvature metric would exceed that of Dark Energy and increase the chance of ripping of space-time unless all matter approaches 3x10^8meters/sec which is "c".

Of course with $8\pi\dfrac{G}{c4}$ *as the gravitational coupling constant approaching the cosmologic constant* ^

the chances of a Big Crunch would approach 50:50. That's 8(pi)G/c^4. It is highly unlikely that a singularity caused by non-natural causes will cause a "Big Crunch" with our level of knowledge and technology. As the Omega point(everything that is learnable had been learned)is reached, the ability to either accidentally or deliberately cause a singularity would increase however with the Omega Point wisdom(knowledge plus experience)should prevent such an occurrence, although according to Quantum Mechanics there is a quantum state where a singularity will occur either accidentally or deliberately. This is delving on the realm of philosophy however.

As mentioned previously, during "a Big Crunch" time will slow down, stop or and reverse time's arrow as one goes from a greater to less entropic state reversing the Second Law of Thermodynamics during this singularity. Time's Arrow will reverse but it will still seems like it's going forward to everything and everybody within the system that the time is reversing in. It would be like saying that his body clock is synchronized for time's arrow to be pointing backwards not forward as Steven Hawking postulated happens in a Big Crunch and what this author agrees with. In this case time will point backwards until it reverts to time=0 at the point of the Big Crunch when space-time expresses infinite curvature as it does before The Big Bang and another Big Bang will follow with a heavily rotated component uncoiling in a swirl with a ballooning expansion as with cosmic inflation and the process will repeat. Again with "Heat Death" in the expansion also progressively decreases as does rotation until everything stops moving and everything freezes. This scenerio would be much more likely if there were no boundaries such as 0 degrees kelvin, "c" the speed of light boundary and spacelessness which can only be breached during singularieis. Sadly we will never live to know because the early stages of a "Big Crunch" may only be be shown by a Doppler Shift which is only slightly less to the infra-red and clocks even with gravity may not slow down a measurable degree because the measuring device is part of what's being measured and heavy masses will also dilate or slow time down which would skew the readings. These phenomena follow the equation $\mathbb{R}a\ b\ c\ d = R\ a\ b\ c - 1/2\ R\ g\ ab/R\ ab\ (1/c^2)$ where space-time is curved inward. Space-time is pulled outward by inflation or with H(the Hubble expansion coefficient)with slight rotation leading to a possible swiss cheese effect in our space-time fabric in which according to Guth and others

14 other universes can form with the same or different physical laws ;however the WOMP doesn't clearly show a swiss cheese effect in space's BMR. In this cast the equation of everything becomes $\mathbb{R}$ a b c d=R a b c+1/2R g ab/R a b times 1/c^2 instead of -1/2 R g a b because space-time is curved outward rather than inward.

Also it is an axiom that each and every intelligent observer must rust his or her perception of the information gathered by the measuring device and that each measurement is based on the Heisenberg Uncetainty Principle which state measurement ranges for everything measured as the measuring device affects what is measured changing it. The is also the information exchanges of relating to "Spooky Action at a Distance" and quantum states or levels changing polarity or information at vast distances. It is also an assumption that the observer exists.

CHAPTER TEN

WHAT OCCURRED BEFORE THE BIG BANG AND WHAT CAUSED IT?

The accepted theory was that time and space began at 10-43 seconds before "The Big Bang" also known as the "Big Bang ex nihilo". This theory opposes Newton's Law "Every action must have an equal but opposite reaction as well as "The Law of Conservation of Energy" which opposes the spontaneous appearance of a quantum bubble out of nothing (including space).

It seems logical that an implosion of another universe could have occurred prior to the explosion of the "Big Bang" in which times arrow was reversed during the implosion of the previous "Big Crunch" at which time instant time almost stopped then reversed time's arrow from backwards to forwards.

There may have been a massive Higgs Energy field which converted a quantum bubble at 10-^33cm to matter according to be variation of energy=mass c^2 and D-0-Branes may have also existed where the zero dimensional state approximated the infinite dimensional state where the infinite dimensions were compactified or curled up with a quantum foam or something else. The multi-verse idea 15 with a n implosion preceding the explosion of "The Big Bang" seems possible.

It is also possible that three universes collided where one was predominantly matter, one predominantly anti-matter and and one a vacuum universe containing energy only. This may be an unlikely event however quantum mechanics states that all permutations will occur including this one. In the anti-matter universe time moves forward and with matter time moves backwards. In the matter universe time moves forward due to positive mass and antimatter time moves backwards due to negative mass. In the "vacuum universe" there is space due to the energy of possibly the Higgs Field and has mass due to the Higgs Bosons which are massive and almost equal matter and anti-matter, however there is slightly more matter than antimatter. As a result of the three way collision, the energy of the Higgs Field of the "Vacuum universe" triggers the implosion of the matter and anti-matter universes triggering a reversal of times arrow and imploding the mass of the antimatter and matter down to a quantum bubble which is approximately string sized which will subsequently explode 10^{-43} seconds later into "The Big Bang" with initially high rotational vector at maximum magnitude just at Planck's Time then slowing at a geometric progression while the expansion or inflation occur with "The Big Bang" moving the rotating 750 billion strings into approximately 6.75×10^{34} erg of energy in a 360 degree of 2π *Radian orb blast with a rotational vector ω and the Hubble Expansion Coeffiicent* $H^{\wedge}$.

Collisions of galaxies occur infrequently but they do occur, as there is a theory that Andromeda is approaching The Milky Way for a collision in several billion or perhaps many many million years, so the collision of universes can and must occur also therefore the term "Universe" is a misnomer but is rather a "Multiverse". Matter universes exist, anti-matter universes must exist because antimatter exists. Pure energy universes must exist with the Higgs Field because the Higgs Field exists. As all of these axioms are true the collision can occur and must occur at time "t" though highly infrequent. An since this scenario will cause the quantum bubble to occur and since a vacuum universe will force an implosion rather an explosion, this hypothesis seems very likely. Also the area between membranes of universes has to exist it can be called or referred to as "ether", which would include D-0-branes which merge with n-branes in a matrix of Quantum Dots. This explains "The Big Bang" or swirl and what precedes it where the Big Bang isn't out of nothing.

From this author's first book "Megaphysics, A New Look at the Universe" it was postulated that two two anisotropic manifolds of ten dimensions each had a singularity with a dimensional reconfiguration of two six-dimensional manifolds and two four dimensional manifolds which resulted in a twetnty dimensional manifold of different configuration than the two initial anisotropic manifolds or surfaces which were unstable. With regard to theories regarding the origin of this universe Michio Kaku's idea that a 10 dimensional syupersymmetric anisotropic universe had a "popped bed sheet" effect with the popped bedsheet being the six dimensional Calabi Yau

Manifold which was string sized and the four dimensional superstring cosmic universe made more sense than the "Big Bang ex nihilo".15In my theory that alternates from the three way collision theory there were two ten dimensional manifolds totaling twenty dimensions and a six dimensional Calabi Yau manifoldor something else possibly related to the Higgs Boson with the Higgs Field acting as "ether" cohabitating a "false vacuum" containing infinite space and energy from the Higgs Field. In this scenario a Higgs' Field Universe, matter universe and antimatter universe didn't have a three way collision nor were there an almost infinite number of multiverses, although there still were an infinite number of dimensions primarily forming D-0-branes where the dimensions were compactified from infinity to zero in a quantum foam or quantum dots in the "ether" with the two Calabi Yau manifolds and two 4 space superstring manifolds forming twenty dimensions with a huge six dimensional Calabi Yau manifold forming the other six dimensions of the non-compactified 26 dimensions of string theory cohabitating the quantum dots of D-0-branes.

The twenty six dimensions indicated by string theory included Ramajian's magic number of nature of 24 as the most stable state of four Calabi Yau Manifolds on the dimension of time and another dimension involving the Higgs Field which incorporates into the other set of four six dimensional Calabi Yau manifolds with time moving backwards or forward depending on the properties of non-compactified space and their near infinite dimensional non-compactified quantum dots. An absolute vacuum developed in an infinitely short period of time(time approaches zero but doesn't reach it)and a tidal wave of near infinite space region engulfed space-time and its contents creating a huge amount of vacuum energy from a singularity according to the equation E=mc^2 which acted as a PRIMER or catalyst either for the Higgs Field to act on the three way collision of the matter, antimatter and Higgs Field Universe to form the quantum bubble or the mass formed torn or ripped the symmetry of the recombined two ten dimensional super symmteric anisotropic universes to form an antimatter quantum bubble a matter quantum bubble both 10-33 cm with one 12 dimensional forming two Calabi Yau manifolds and two 8 dimensional forming two quantum bubbles which each contain 4 space with opposite spin2 vectors ;one clockwise one counterclockwise forming the rotational vectors of the initial 10-43 seconds before the "Big Bang" or swirl. The eight dimensional string bubbles had a crushing Ricci Tensor which was opposed by the anti-particle anti-particle repulsion twisting it into a double torus configuration which is more stable than the 8 dimensional complex causing rotation to occur greater and greater at the ends of the bubble(s)where the rotations were in opposite directions(differential spin)until the double torus acted like a pretzel and finally divided in in the middle of the two four dimensional quantum bubbles rotating in opposite directions with most of the mass occurring at the ends or outer edges and finally reaching a velocity to exceed quantum gravity of the two more stable isotropic four dimensional universes leaving two 6 dimensional string universes. The bulk of the mass spun out like a potter's wheel toward infinite length but remained string sized with Calabi Yau Manifolds at 10-33 cm. Whichever of these ideas is the simplest according to Ockham's Razor (Occum's Razor) is the most likely but again with quantum mechanics there is a probability that both occurred. In any case," The Big Bang ex nihilio" doesn't seem likely in terms of Newton's Third Law and The Law of Conservation of Energy and Momentum.

CHAPTER ELEVEN

DOES NOTHING EXIST?

By definition nothing means non-existance. Mathematically space-time/space-time is described by the equation $-1/2e^{\wedge}$in $\cot\theta$ *divided by* $-\dfrac{1}{2e^{i}n}$ $\cot\theta$. *In the expression i n cot theta where theta is π radians you get i∞cotθ or-1(i∞)if θ=π -radians* $-\dfrac{1}{2e^{\infty}\, for}n$ *for) n dimensions where n→∞-$\dfrac{1}{2}$e to the infinity power is=∞. So as space-time is-$\dfrac{1}{2e}$-in cotπ is ∞. The expression $\dfrac{\infty}{\infty}$ is everything except 0(zero). So as the-$\dfrac{1}{2's}$ cancel out e-i n $\cot\theta$ where n →∞ gives ∞ in the numerator and denominator plus THE IDENTITIY POSTULATE IS MET SO WITH THE INFINITE*

INFINITE SPACE-TIME IF THERE ARE INFINITE DIMENSIONS APPROACHED IN THE MULTIVERSE. As space-time is never stationary except in a spaceless absolute vacuum due to space-time's motion(eventhough it goes below Planck Length at the Event Horizon of a black hole or pre-Big Bang or post Big Crunch)the infinite sum of space-time ∞ *and must be* ∞. Π(*eigenstates 0 to ∞ of a*1→*an where a=space-time diverges to* ∞.$\dfrac{Again\infty}{\infty}$ *is everything except* 0. *If the expression was* $\dfrac{0}{0}$ *then 0 would have been included*

Included. But as zero spacetime is not the infinite sum of spacetime the former MUST BE THE CASE. The exception would be if all space-time moves in equal and opposite direction with the same magnitude always.. As the Riemann Metric for space-time with its 256 permutations equal zero net and the Bianchi Identity cancels out space-time in equal and opposite directions in many covariant and contravariant tensors there is an argument that spacetime(net infinite sum)/spacetime(net infinite sum) is 0/0 which includes the null set or spacelessness. This is true as 0/0 includes 0 as a solution. As the meat of the question is does all space-time as an infinite sum of the metric of R a b c d equal zero. According to Riemann it does. This indicate the mirror space-time manifolds of -1/2 e ^I n cot theta and -1/2 e –in cot theta as the left and right handed components of space-time in the Big Swirl and Anti-Swirl mentioned in this author's first book "Megaphysics, A New Look at the Universe". Ths is true in the n-dimensional state where n approaches infinity with regard to dimensions. Although space is a container by definition and a container can exist without contents this seems to go contrary to logic although Einstein apparently stated that mass creates space but this precluded "massless" energy fields with photons which while they have a positive moving mass have a zero resting mass according to sources. Spacelessness is a boundary which if breached can implode a universe and possibly the entire space-time continuum reverted to a Higgs Field of background energy and quantum dots which are infinite non-compactified dimensions in a strict lattice formation as described by the 2 dimensional lattice equation based on the critical exponent and the act of bosons on a circle which is a compactified form of type IIa string theory or M Theory. The spherical model 17 if solvable in the presence of a field such as the Higgs Field. The spin takes on real and imaginary values and interacts with all the spins of the quantum dot lattice. It is subject to the constraint $\Sigma a=1$ *to N* σ^2 *where σ ais the first eigenstate of the sum of sigma squared=N. Couples that with the thermodynamic limitof G(Gibbs Free Energy)=* $-kT/2\pi \int_0^x F(\theta)d\theta$ *based on the Boltzman Equation of States of matter interacting with energy* and theta becomes infinitely small we develop from the two dimensional Ising Model a two dimensional statistical model where a parition function $Z = \acute{O}ne[-\dfrac{E(n)}{kT}]$ *where E(n)is the energy of the nth state k is the Boltzman constant and T represents the temperature* down to zero degrees kelvin. The free energy F=-kT ln Z at criticality correlation functions between spins σi *andσj develop a metric g ij. The expression g ij=<$\sigma i\sigma j$>-<σi><σj> depending on the distance "x" or "r" spearating the states. The correlation length E becomes* ∞

At a phase transition from one state to another and at large distances where x or r approaches infinity g i j approaches x-t*e-x/ ε(.Δ=conformal weight) Thus g ij=x^d+2-η=x^{-2Δ} where these are approximations. While ηis the critical exponentof the field the energy operator ε.* 16. Phase transitions on a quantum dot level approach the Ising Spherical Model 18 energy operatorat ε as a product ogf two fields whereby ε n=σnσn+1 then it follows tha t<εnε0>equals x(any observable or expectation value $^{-2\left(d-\frac{1}{v}\right)}$ has an infinite correlation length at a critical temperaturecritical temperaorat a critical temperature which was already mentioned. The interaction of a free boson or Higgs Field on a quantum dot matrix may follow the Ising Spherical Model as the circle is compactified type II a sting theory. If spaceless ness occurs a fracture will occur in the quantum dot lattice upsetting the spins $\sigma 2^2$ *breaking the continuum N relating to the sum of spins. This state is in the zero external* or fermionic (vacuum state). This is shown by the Grassmann Oddvariable $\psi\text{-}n^{2|0}$>*with the nth Fermionic Oscillator trace being* $\psi\text{-}n|0$>*and vacuum state.*

This describes the quantum dot lattice in terms of two dimensional open or closed strings using the Ising Model 19. As the Fermionic vacuum state and quantum dots are not nothing and the spacelessness state actually can be subdivided down toward infinity where an infinite number of dimensions equals zero dimensions as in the D-0-Brane spacelessness doesn't exist.

CHAPTER 11 continued

This describes zero external field limit which approximates the limit where fracture of the lattice of quantum dots may occur localizing spacelessness.

Basically space appears to be infinite in magnitude and direction but space-time appears to be infinite in magnitude but finite netting zero in direction. These observations may relate to times arrow moving forward or backward. At the speed of light boundary "c" space-time approaches zero as time dilates toward infinity. At the event horizon of any black hole space-time approaches zero as time dilates toward infinity. Finally, the question of whether nothing exists or not depends on whether space is infinite with an infinite sum of infinity with regard to magnitude and direction or if space is 0(zero) with regard to direction while it may be infinite with respect to magnitude which would include what this author calls "an infinite space vacuum" or pseudo vacuum if energy exists in infinite space as compared to absolute spacelessness.

Based on the Relativistic form of the Equation of Everything $\mathbb{R}$ a b c d=R a b c-1/2R g a b/R a b as the mass reflected by R a b goes up R g ab the space-time curvature variant from the metric g a b representing the effect of gravity on flat or Mintkowski Space R a b c can curve space from an asymptotic flatness to an infinitely curved point at the event horizon of a black hole making the expression 0=0/R ab where R a b represents the inertial mass. Based on this space-time can exist at an infinitely or near infinitely small size with infinite curvature. This still isn't spacelessness. It is the opinion of this author that spacelessness is a boundary or interface which requires a force which may equal or exceed a "Big Crunch" implosion to breech however as an asymptotic boundary it may not be breechable even with the implosion of a "Big Crunch". The term impossible means "that which is beyond the comprehension of mankind" as quantum mechanics says every permutation that can occur does occur therefore spacelessness if and only if a permutation of space and space-time can and will occur.

The equation of a circle arc length(s)=$r\theta$ where r=radius if the arc length approaches zero and the angle θ *approaches zero they are ASYMPTOTIC FUNCTIONS* AND AS ONE SECOND OF ARC CAN BE DIVIDED DOWN AN INFINITE NUMBER OF TIMES WITHOUT REACHING ZERO THEN THE ARC LENGTH REPRESENTING THE MAGNITUDE OF SPACE-TIME WOULD BE DIVISIBLE DOWN AN INFINITE NUMBER OF TIMES AND NEVER REACH ZERO. THIS IS WHERE S=SPACE-TIME AND THETA IS THE SPACE-TIME CURVATURE METRIC OF ANY AND ALL "r". This corollary assumes the axiom that IIa string theory compactifies to be a circle which is a documented fact. Therefore the tensors of the fourth degree whose magnitudes are additive whether positive or negative and whose directions are positive or negative in a near infinite number of intersections of osculating planes then net space-time can be zero but due to LOCALITY zero is virtually never reached or measured by intelligent observers. When a measuring device is local and part of being measured the observational viewpoint of zero space is beyond the comprehension of mankind, however not necessarily beyond the observation of other intelligent observers which are not local. THIS IS IMPORTANT. The reason why nothing doesn't exist is because based on the collective cognitive abilities and information gathered by a hypothetical or real observer or observers is being the comprehension of that hypothetical or real intelligent observer or observers. Mathematically the null set has been derived and used many times by mathematicians and physicists and is therefore considered a real finding in nature but mathematical computation and comprehension of a phenomenon are two different and distinct things.

CHAPTER TWELVE

THE ANTHROPIC PRINCIPLE

The anthropic principle states one and only one thing. We are here. We exist. Based on this idea there must be a specific combination of events to make possible intelligent life. This series of events includes the formation of amino acids, RNA, DNA, oxygen, water and an environment which is totally conducive with the formation of intelligent life. This in essence is the sum total of the anthropic principle but to amplify it there are ideas that without "intelligent observers" an event cannot be proven to have occurred. The question "When a tree falls in the forest and there's nobody around to hear it does it make a noise?" The answer isn't obvious. One would think using judgment that the answer is yes as the Physics of a tree falling with an atmosphere would create sound waves which follow the Doppler Effect. The fact that there's nobody around to hear it doesn't affect the formation of the Doppler Waves creating sound. It is true that there would be no body around to hear it or measure it with a detector but that doesn't mean the sound doesn't exist. In Quantum Mechanics an intelligent observer affects measurements within an uncertainty range and there are no absolute measurements only ranges which is why measurements involve an expectation value<> and probabilities $\|x\|^\wedge 2$ rather than absolute measurements. This is because the measuring device affects what is measured and LOCALITY EXISTS IN SPECIFIC STATES where local phenomena are affected by everything in that region including measuring devices. However while it's true that the sound made by a falling tree in a forest will and must be affected by an intelligent observer being present. the effect should be miniscule. ALSO IF AN INTELLIGENT OBSERVER HAPPENS TO BE DEAF OR DEAF AND BLIND AND IS PRESENT IN THE FOREST WHEN THE TREE FALLS, A NOISE IS STILL MADE. Therefore "common sense" says the answer must be yes, but it is affected by the intelligent observer being present. This implies that all events in space-time which occurred before the formation of life or intelligent life did indeed occur, as the converse is unlikely. If there was no intelligent observers during the first 12 billion years after the "Big Bang" how could the Milky Way" and the planet earth have been formed. Without the milky way and the planet earth being formed, how could we exist as intelligent observers? The converse of this case is if other intelligent observers existed going back to the Big Bang. Who or what are the intelligent observers that existed during "The Big Bang"? The answers go into the purview of religion which is not the objective of this book although The Higgs Field and Higgs Boson have been postulated as "The God Particle" and the Higgs Boson was isolated in Cern, Switzerland by the Hadron Collider. Prior to this, this author had some evidence that the non-isolated tachyon may have been the God Particle as it acted like negative mass rather than positive mass and appeared to go backwards in time where times arrow was still moving forward rather than backwards. Also the tachyon apparent broke other rules of physics such as exceeding "c". As the tachyon apparently becomes a boson at v<c when it is no longer a tachyon, it could still be a property of the massive Higgs Bosonic Field.

Is There Life Elsewhere?

The laws of probability say the answer to this question is undoubtedly "yes". Is there intelligent life elsewhere? The answer to this question according to probability is also" yes." Is there local intelligent life elsewhere which is not related or branch to our life forms? The answer to that by probability is "no". It is likely that on moons such as Io or distant newly detected moons in other solar systems there is an environment to support life, but the likelihood that this life is either vegatable, fungal, bacterial or viral. Animal life is so complex in its intricacies that to have a celestial body with animal life(even simple animal life) would have to be large distances from the earth's detecting devices. It is certainly possible that there is intelligent life in The Milky Way, however the frequency of this occurrence would be so low that this author might be talking about fewer than five celestial bodies; while for bacterial or fungal life or vegetation the number of celestial bodies may be in the thousands. It is also possible that there are no celestial bodies with intelligent life outside the earth or perhaps only one in this vast galaxy. It may take many years to answer these questions if at all.

Was there ever life on Mars?

Missions to Mars include the Mariner series, Viking 1 and 2, Mars Observer, Global Surveyer, Pathfinder, Climate Orbiter, Polar Lander, Deep Space 2, and Phoenix. Also 2201 Mars Odyssey, Exploration Royers, Mars Express and Reconnaissance Orbiter, Mars Science Laboratory and Maven. Mars may have water and in the past an atomosphere which means that sub terranium life may have in the past existed or may exist now under ice. Gathering information about "The Red Planet" takes years and may or may not give a direct rather than indirect answer. Bacterial, fungal or viral life is probably more likely than plant or animal life. Certainly to have photosynthesis it would be difficult to occur beneath the surface of the red planet and while it's possible that fossils may exist it still be many years to get the answers. Underground and bacteria or viruses it may exist or may not.

Moons such as Jupiter's Io were also been considered life friendly due to volcanism. It is unlikely that our civilization will have concrete answers in the near future. If one trust's one's senses and perception we are here. That is a fact. Do others exist besides the individual that is perceiving the environment your brain says "yes' to you but is the answer yes. We make assumptions about our environment and then follow its rules. Would physics exist without life ? If there's life in this universe is the universe alive? We have obvious answers to these questions, but are they correct?

These answers delve into philosophy. Can the reader prove he or she cohabits this universe with others? By the assumption that the reader isn't the author the reader can convince himself or herself that it's true. Physics SHOULD exist with or without life but can't be proven as it takes life to prove it. If there's life in this universe and the life is part of the universe then the universe is alive, at least in part. Do others exist besides the reader? Probably because the reader's brain can prove it to the reader. Is there more to this universe than what is measured and perceived? This is unknown and may remain unknown as a cell cannot perceive a body it inhabits or if it is independent. We are here can be amended to "I am here" or "you are here "being the reader.

The Anthropic Principle puts constraints on the definition of life, intelligent or not. It implies that life requires an environment conducive of life which can require food(ingestion or photosynthesis), respiration, have reproduction, locomotion(very slight in plant life)excretion, and can be born and die. It states that conditions that form "us" require D.N.A., oxygen, ambient temperatures and water as well as a P.H. which will not destroy important tissue. It deems it unlikely therefore that the entire universe is alive Veneziano(1993)18. Although galaxies rotate and in dilated time can act as dynamos producing energy, black holes can act as an excretory function such as Mallipigian Tubules and "The Big Bang" can act similar to fertilization. This supposedly preposterous hypothesis would answer the question "What is the meaning of life" besides perpertuating life would show that every microcosm of life have a function in pertpetuating the whole, although in terms of dilated time an organism such as us could live only an instant in the scheme of things just as a bacterium would live only a few days compared to a human being. Viruses perpetuate viruses, bacteria perpetuate bacteria, plants perpetuate plants and animals perpetuate animals. The food cycle perpetuates life on earth and the environment. Intelligent observers may perpetuate life on this planet by preventing global catastrophes such as an asteroid strike or perpetuate distruction such as with a cobalt bomb or H Bomb. How would life on the planet earth perpetuate a "living unverse' in dilated time. The answer to this question is unknown and may never be known to us;but that doesn't mean it automatically means "it can't".

Existentialism states that "life is meaningless". Is this true? I life has a function in perpertuating life on a larger scale then the statement would be false but on a smaller scale life does perpetuate life.or attempts to. We are trying to prove that "we are not alone in this universe" and while we are closer than in the 1960's or 1970's we have not proven it yet as of this date in early 2015.

Paranormal phenomena are used to determine that life exists after death and while there is extensive anecdotal evidence of such life, it is difficult to prove empirically with scientific measurement except for a drop in kvp and temperature around "cold spots" in the environment where orbs exist and paranormal phenomena have been reported.

CHAPTER THIRTEEN

QUANTUM MECHANICS

It is impossible to absolutely measure or localize anything measurable in space-time. Everything measured must be in range based on the probability of any point particle $|(r,t)|^2$ and is described by the wave function $\psi(r,t)$. This is based on the Heisenberg Uncertainty Principle and defines the Schrodinger Equation.

Assuming the velocity of space-time is *πc or the product ofπ and c or the speed of light and that the sum total of all Riemann Forces and energy is zero*

and energy is zero then the wave function of any point particle must travel at less than *πc which is the speed of space-time. Each point particle has a deBroglie wavelength which is* $\lambda = \hbar \dfrac{}{mc}$ *so there are wave properties to any matter with mass although photons have a resting mass*

mass approaching zero while there is a measurable mass while photons are in motion. Before discussing this topic one must prove that space-time travels at pi(c). In a circle which is the compactified form of type IIa string theory the circumference describing space-time approaches infinity with reaching it and the diameter involving the sum total of all Riemann forces approach zero(0)with a domain of $\infty$to∞. $-c/d = \dfrac{\infty}{0}$ *with a domain of-$\infty$to∞ including the null set of 0. The denominator is the null set {0} where space-time is a open dinvergent* set *and* 0 is a closed convergent set. $\dfrac{\infty}{\infty} =$ *everything except* 0 *and* $\dfrac{0}{0}$ is *everything including* 0. $\pi = \dfrac{\frac{22}{7} \ and \ \frac{1}{\pi} \ is \ \frac{7}{22} \ so \infty}{0}$ *times* $\dfrac{0}{\infty}$ *is* 0. *Assuming that space-time=circumference and diameter is mass and energy* $= Rab\left(\dfrac{1}{c2}\right)$ *of the metric g a bdoing the curving on R a b c which is Mintkowski Space or Rimenann*

Space-time. Again θ is the space-time curvature variant $R \ a \ b \ c - \dfrac{1}{2} R$ g ab where R g ab is the curvature variant from metric g a b.$\pi$is the circumference of a $\dfrac{circle}{diameter}$ of a circle

Circle/diameter.s=r(theta) where s= space-time, r is the sum total of Riemann forces and theta is the space-time curvature metric. The circumference of a circle is without beginning or ending therefore converges to two pi radians or 360 degrees as an asymptotic function without reaching it. Therefore the absolute measurement of a circumference is is not possible but only an approximation of a convergent function to two pi radians or 360 degrees. Therefore the space-time continuum as the circumference being in the numerator/ total Riemann forces in the denominator with +forces/-forces/matter and anti-matter causes curvature and reciprocal curvature of space-time between the +forces and – forces causing the swirl and anti swirl of The Big Bang". Space-time travels at pi times the speed of light as it must do so to exceed the Riemann forces to "outrun them". Again the circumference of a circle defines a continuum as it is without a beginning or ending and converges to 2 pi radians therefore going to infinity. The diameter isn't bounded as the matter and energy can never reach the circumference as it can't exceed "c" or the speed of light boundary while space-time travels at v>c by pi. Without boundaries for matter and energy the diameter can travel outward a near infinite distance without reaching infinity. As a result the equation space-time/mass can never actually reach pi or 22/7 which is why pi goes on to infinity without reaching a definite decimal termination 3.1415928.... If the diameter of Riemann forces was bounded space-time would be slowed down to v<c or v=c in a "Big Crunch" and everything would be compactified to pi radians or 180 degrees which is the trajectory of "The Big Bang". This is when the sum of Riemann Forces=1 so the circumference/diameter becomes 22/7 divided by 1. Aagin above v=c the circumference travels at v=pi© while the diameter travels at v<c including light and photons which do not travel at c but just below it as "c" is a boundary to everything except space-time and tachyons. In this universe the Riemann forces must converge in a "Big Crunch" and divergent space-time will also converge from infinity to 22/7 or pi radians while the denominator or Riemann forces converge to 1 yielding pi radians at the central core or the quantum bubble of the "Big Crunch" Times arrow must also point backwards in this case as infinite space-time is converging to pi radians. Finally, you can not absolutely measure the circumference of a circle (as mentioned previously)as it has no beginning or end so the number of rotations of the circumference must be n where n approaches infinity or infinite

number of turns of 2 (pi)radians each as well as an infinite number of subdivisions. Therefore infinity/0 approaches 22/7 in a bounded convergent state which is the circumference of a circle/diameter that is where the domain goes from –infinity to +infinity to $0<\theta <2\pi$ *radians causing the constant 22/7.* It also causes the velocity of the space-time continuum to travel at πc or $n\pi c$ *where n is any real or imaginary number.*

The Schrodinger Equation describes the wave function of a point particle $\psi(r,t)$ *in a spacial state of three coordinates at time t. The probability density is d)(r,t)ord3r*

At time t for point r.dP(r,t)=C| $\psi(r,t)$|2d3r *where* $\int dP(r,t)=1$ *and* $\psi(r,t)=\int |\psi(r,t)|^{2d^{3r}}$ *with the normalization constant C where* 1/ $\int ||(\psi(r,t)|^2d^3r$ *has c=1 as a normalized wave function and* $\psi(r,t)$*is continuous everywhere. So the Schrodinger*

CHAPTER THIRTEEN

Schrodinger Equation is $i\hbar\dfrac{\partial\psi(r,t)}{\partial t} = -\dfrac{\hbar^2}{2m}\nabla^{2\psi(r,t)} + V(r,t)\psi(r,t)$

where ∇^2 is the LaPlacean Operator or partial differential equation in the x,y,z coordinate system

Ordinate system of the function being operated upon where it is a second degree partial differential equation. The $\dfrac{\hbar^2}{2m}$ *operates on the LaPlacean Operator* $\dfrac{\partial^2}{dx^2} + \dfrac{\partial^2}{dy^2} + \dfrac{\partial^2}{dz^2}$ *to form the Hamiltonian Operator whereħ is Planck's Constant or 6.63x10^-34.joule-sec.*

Planck Mass is the smallest mass that can be occupied by quanta and is described as *ħc/8πG)^1/2 or simplified as(ħc/G)^1/2. This was used as the inertial mass of a quanta when applied to the probability distribution of* a point particle in n-dimensional space with space-time as -1/2e^-in cot θ *in the Equation of Everything where n is the number of dimensions and θ is the angle of trajectory* trajectory of "The Big Bang" or π radians with space-time traveling with a magntude of πc and direction of rotation and expansion with rotation decreasing exponentially and expansion increasing due to effect of Dark Energy from the mutual repulsion of anti-particles compressed to 7.5x10^11 strings in the quantum bubble. There were 6.75 x10^34 erg force unwinding the extreme rotation with the antimatter repulsion of Dark Energy all initiating at Planck Time 10-43seconds.

The presence of an intelligent observer has changed scientific measurements. It has been determined that the position of an electron or positron was altered when it was measured from where it was mathematically determined to be. The presence of a measuring device affects what is being measured and electron are postulated as clouds which dissipate over distances but still may exchange information such as spin over long distances. Photon Pairing was incorporated with "Spooky Action at a Distance" where information was exchanged over vast distances between paired particles. Also, the presence of a subatomic particle has been shown to be in a region of space with no observer and not there when an observer was observing it. On an electron cloud this is explainable as the cloud can be huge and exchange information with another electron cloud. Perhaps the same can occur with photons as light or electromagnetic radiation has wave and particle properties. Also space-time approaches zero with out reaching it near "c" and since electromagnetic radiation travels at near "c" and electrons travel at approximately 2/3c so space-time is narrowing at these speeds, and with reduced space-time vast distances can be covered by photons and electron clouds although the information exchange in Spooky Action might actually be greater than "c". As space-time travels at πc *the motion of space-time may carry the information of these paired particles with it or it may relate to time dilation at or near "c"* dilation at or near "c". The presence of Schrodinger's cat in a box has been shown to be affected by the presence of an observer. Everything measured is a region or range with a probability distribution of the occurrence being at the top of the probability curve with less likely measurements of the occurrence at the ends of the probability curve indicating there is no clear ABSOLUTE MEASUREMENT only probable ranges of measurements. Louis de Broglie stated in 1924 that there is an associated wavelength and therefore frequency for each and every mass making a wave function of point particle r at time t result in a measurement range as in the Schrodinger Equation. Werner Heisenberg stated that there was uncertainty in each and every measurement resulting also in a range or probability distribution for the true or absolute value.

When the Boltzman Equation is made for different states of matter at specific temperatures such as 0 degrees centigrade for the change of ice to water that temperature is fixed, so how can it be a range? If the heat of vaporization of liquid water to water vapor at 100 degrees centigrade is based on mc ΔT+*mlf or mlv how can the measurement not be exact when the calories are expended to vaporize water or liquify ice?*

Mlf is the mass times the latent heat of freezing. Mlv is the latent heat of vaporization times the mass. This c is specific heat and T is the temperature generally in degrees kelvin. Q is the number of calories or kilocalories of heat required to raise the temperature nd to change the state of the substance as in Q=mcΔT. These constants result from thousands of measurements each with a probability range and the median reading results in the constant so the Heisenburg Uncertainty Principle still holds. Constants in nature result from thousands of measurements with probability distribution curves not unlike quantitative analysis measuring percent purity of compounds or elements. Even Planck's Constant would be the result of thousands of measurements. The same should apply to the speed of light constant "c" and the gravitational constant "G" as well as the gravitational coupling constant "k". Space-time curvature measurements were done under rigorous conditions of a solar eclipse and had to be done multiple times although this author does not know how much of the Heisenberg Uncertainty was involved in these measurements. It is difficult to show how Relativistic changes based on measurements rather than tensor calculus was affected by the Heisenburg Uncertainty and it makes this author wonder Albert Einstein felt about the Heisenberg Uncertainty Principle. Even the Lorenzian Transformations would have to be subject to the Heisenburg Uncertainty Principle with regard to testing the equations with scientific measurement. In statistics the p value has to do with percentage error in the measurements and generally is incorporated in any and all statistical studies. Precision of properties or states of matter and energy involves the clustering of measurements in a statistical range of an answer or measurement. Does precision imply accuracy? Reproducibility is a property of measurements that are accurate. Precision implies reproducibility but it must be measured under different conditions and probably by more than one measurer. The measurement of 22.4 liters of an ideal gas at standard temperature and pressure in the calculation of a mol of an ideal gas must have been from thousand of measurements as with the ideal gas law PV=nRT where R is the combined gas constant 0.0821 T is the temperature in degrees kelvin P=pressure V=volume and n relates to the number at mols or millimols of the gas which can be converted using the factor label method to Standard Temperature and Pressure. The same applies to Boyle's and Charles' Law.

The expectation value< > is the, measured mean or median value of what's expected in the measurements and the wave function $\psi(r.t)$ *relates to the wave property of a point particle r with respect to time t.*

The Quantum Mechanics form of the Equation of Everything or Space-time=Space/mass($1/c^2$) is based on the expression Planck Mass($\psi|r,t|$ *where* $\psi(r,t) = i\hbar\dfrac{\partial\psi(r,t)}{\partial t} = -\dfrac{\hbar^2}{2m\nabla^{2\psi(r,t)}} + V(r,t)\psi(r,t)$ *according to Schrodinger's Equation. The Hamiltonian Operator is used for weak*

Perturbations or quantum fluctuations. so $\Psi(r,t)dn^r\nabla^n/\sqrt{}\,\hbar\dfrac{c}{8\pi G} + \mathcal{H}(a\to n)$ eigenstates of energy $\left(|r,t|\right)^{2d^{nr\wedge n}} = \psi(r,t)d^{nr\nabla^n} - \dfrac{1}{2}e^{-i}n\,\cot\theta$ *where* $\psi(r,t)$ *is the wave function,* $d^{nr\nabla^n}$ *is the number of dimensions of point r in n dimensions and*

Dimensions and $c^2(hc/8(pi)G)^{1/2}$ is the Grand Unification Energy GUT for Planck Mass. H(a to n eigenstates $d^\wedge$ nr ∇^2 *is the Hamiltonian for n eigenstates of energy of the probability amplitudes of point particle*

R at any time t or it's wave function.$|(r,t)|^2$ is the probability of r,t in n dimensional space while the Hamiltonian H(a to n eigenstates) of the probability amplitude or wave function of r at time t covers weak perturbations or quantum fluctuations. As n dimensions approach a very large number due to the infinite number of intersections of a subdivided second of arc in rotation and expansion producing osculating planes which are non-parallel -1/2 e –I n cot theta approaches -1/2 e - ∞*power or 0 which solves the Equation of Everything for Quantum Mechanics with regard*

CHAPTER THIRTEEN CONTINUED

to Planck Mass. In the n-dimensoinal state quanta with a mass of Planck's Mass acting as a point particle at time t with a wave function obeys the equation ℝ a b c d=Ra b c -1/2 R g ab/R ab with the Hamiltonian Operator acting upon a to n eigenstates of energy to include weak perturbations from quantum fluctuations.

QUANTUM FIELD THEORY AND GROUP THEORY

Quantum Field Theory is the incorporation of Group Theory into Quantum Mechanics and constraints are made into quantum field theory such that all fields are irreducible representations of a series of groups with iso -spins as well as Lorenzian Groups and Poincare Groups and that the theory be UNITARY WITH A RENORMALIZABLE ACTION WHICH DOESN'T VARY IN THESE GROUPINGS. The action(S) must be causal such as space-time being curved by mass as the effect of gravity and there must be consistency with these re-normalizable actions(s).19

Space-time symmteries include any and all Lorenzian and Poincare groups or groupings. These symmteries are non-compact whereas the endpoints are not included in the range such as with asymptotic functions such as space-time approaching zero at or near the event horizon of a black hole but never reaching zero space-time or with matter approaching the boundary of "c" or the speed of light but never reaching it ; although space-time "pinches off" at c traveling above "c" by *cπ*.

As space-time approaches "c" from πc *to (π-ϵ) c during a "Big Crunch"* it will narrow down towards the asymptotic function as it would at just ϵ *below c as π→1. In this case the directional vectors would have the time coordinate facing opposite in direct* direction but equal in magnitude as with a geometric expansion pointing times arrow away from forward initially ever so slightly and then more and more as space-time almost (but never does) pinch off at the speed of light boundary.

Internal symmetries include groups or groupings of point particles that intermix such as with the Strong Force's quarks mixing different flavors of quarks among themselves. Internal symmteries rotate fields and wave functions relating to point particles in what has been coined "Isotopic Space". The groupings are compact with definable endpoints unlike space-time symmteries. The rotational subgroup can occur as in a circle between zero and 2π *radians where the limits of integration are included albeit different from the case where space-time has limits which are zero to 2π radians in the circle which was compactified by typeIIa*

String theory as the limits are "outrunning" the diameter which contain matter, energy and any other unnamed quantities within that subset of the set containing space-time. The "outrunning phenomenon" states that the diameter of a flat orb or circle will approach but not reach the circumference as long as the rate or expansion and rotation of the circumference travels faster than the diameter and when the circumference slows to the velocity of the diameter the expression becomes 22/7 or π. *Internal symmetries are either global(relating to the diameter of the flat orb and not*

Space-time) or local whereby local symmetry subgroups vary with each point particle in 4- space.

Super symmetry incorporates space-time and internal symmetries including local symmetry.

Group Theory states that a set A with a fixed binary operation is called a groupoid (.). A groupoid can be associative if a1(.)(a2(.)a3)=(a1 (.)a2)(.)a3 for all a1, a2, a3, in A. An associative groupoid is called a semi-group. An Isomorphism is when elements of two groupoids are the same except for the names of the elements. Homomorphisms are defined in group theory as an element from a Groupoid (G, α) into a groupoid (H, β) is a mapping $\theta{:}G{\rightarrow}H$ *such that the condition* $((g1,g2)\propto)\theta{=}(g1\theta,g2\theta)\beta$ *for any and all* g1,g2 in G. θ *preserved multiplication So a homomorphism of (G,..) and (H,.) in the mapping of* θ.

The mapping of the sum total of Riemann forces upon space-time $R{\leftarrow}\theta$ *where the circumference is describing space-time has a mapping of groupoids ($\sum$eigenstates of quanta as energy levels a$\rightarrow$n) wherethe eigenstates are homomorphic and isomorphic with type IIa String Theory and* Yang Mills GUT or 10^19 Giga electron volts. which describes the flat diameter when space-time decelerates in direction to "c" from πc *and shrinks in magnitude to* $\frac{22}{7}$ *or π radians at or the speed of light boundary.*

Isomorphisms can be applied to space-time with a swirl and anti-swirl where the direction is opposite but magnitude is identical(only if space-time has a mass which it would to provide an inertial moment at "c" to prevent the Riemann forces from breaching "c" due to the near infinite inertial mass of the Riemann forces at just below "c" and is weakening the "fabric" of space-time at "c" which eventually may tear space-time just before a "Big Crunch" when space-time has shrunk from πc to *just above c resulting the* $\frac{circumference}{diameter}$ =22/7.)

The Dirac Equation describes $i\frac{\partial \psi}{\partial t} = -i\alpha$ *(αas a vector).* $\frac{\partial \psi}{\partial x (asavector)} + \beta m\psi$. *There is a relativistic equation that is a first order in the time derivative so that it forms a non-non-negative probability density.* The equation must be Lorenzian covariant and be first order such that space and time are treated equally.

ψ must be spin up and spin down and α andβ are matricies with ψ as a column vector. In order to show Lorenzian Covariance the equation must be treated as a matrix equation. The isospins(spin up and spin down)must fit a 2X2 matrix with regard to *α and β* and space and time must be incorporated into mass in the relativistic equation. Probability density must be positive as negative probability density is the incidence of non-occurrence of events or the probability that events with regard to the Dirac Equation will NOT OCCUR.

As space-time is curved by mass via gravity the element of θ *incorportated by the distributuive law into each function of the subsets of the interaction* of Riemann Forces (that include mass including weak perturbations) Riemann Forces should be homomorphic just as space-time should be isomorphic. In this case θ *is the space-time curvature metric of Riemann Forces on space-time.*

In Quantum Field theory the values of upspin and downspin revealed by ψ *must satisfy the energy momentum variable for a free point particle or its wave function's Free* $E = \dfrac{i\partial}{\partial t}$ *and momentum* $p = -\dfrac{i\partial}{\partial t}$ *where the components of ψ must satisfy the Klein-Gordon Equation regarding relativistic covariance and Lorenzian Covariance with* regard to the Lorenzian Transformations regarding time dilation. The resulting Dirac Equation is $-i\dfrac{\partial \psi}{\partial t \gamma^0} = \dfrac{i\partial \psi}{\partial x}.\gamma + \psi m$ *where* $(i\partial - M)\psi = 0$ *where the partial derivative moves backwards. WITH REGARD TO SPINORS The* Dirac Equation $(i\partial - M)\psi = 0$ *WHERE THE DERIVATIVE MOVES TO THE LEFT SIDE OF 2 THE EQUATION RATHER* THAN THE RIGHT. M is the mass of the spin up and spin down (Spinors) and dealing with fermionic curvature of space-time the action $S[\psi^\wedge{}_-, \psi] = \int d^\wedge 4x L = \int d^\wedge 4x \psi(x)(i\partial - M)\psi(x)$ *where the field momentum conjugate is* $\partial L/\partial \psi = i\psi^\wedge*$ *The Field Momementum Conjugate equals π for field momentum L/ψ as π is the relationship of the circumference of a circle divided by it's diameter* the expansion of space-time with rotational vector divided by the sum of all Riemann Forces gives the Field Momentum Conjugate whereby space-time is expanding at a greater rate than Riemann Forces are expanding and the Riemann Forces are acting with the infinite momentum tensor R on space-time which is expanding at π c *where the Lorenzian Transformation for infinite mass at just below the speed of light boundary is reversed to the left og=*

dary is reversed to the left extreme of the Dirac Equation causing the extreme inertia (R ab) at "c" preventing inertial mass from exceeding or actually hitting "c".

As supergravity was already briefly discussed, it may be redundant to show the existence of the 11th dimension however gauge symmetry groups in Yang Mills Classifications will be discussed later.

Getting back to Quantum Mechanics this author will now discuss will discuss probability density and probability current. The probability density is based on dP(r,t)= ρ(r,t) d^{3r} *where the probability of a point particle in an infanitesimal volume of* d^{3r} *located at* R at time t exists as the above equation. Remember that the probability of a normalized wave function is $\rho(r,t)=|\Psi(r,t)|^2$. *Also the probability dp(r,t)over all space remains constant at all times.*

The Law of Conservation of Probability in a local phenomenon is $\frac{\partial \rho(r,t)}{\partial t}+\nabla.J(r,t)=0.J(r,t)$ is the probability current and the above is the continuity equation. The probability

Current is J(r,t)= $\hbar/2mi\left[\Psi*(\nabla\Psi)-\Psi(\nabla\Psi)*=\frac{1}{m}\text{Re}\left[\Psi*\left(\hbar\frac{}{i\nabla\Psi}\right)\right]\right]$ The Reflection Coefficient R=|JR/J I| where J Ris a reflected wave of probability current and J Iis the incoming wave of probability current where two regions of space are separated by a potential energy stem or barrier. The transmission coefficient is T=|J T/J I| 20

The above shows that probability density must be positive as negative probability for occurrence is the same as positive probability of non-occurrence. The above helps solve the "Particle in a box" problem while J refers to the current and J I refers to the intensity of the current.

The Harmonic Oscillator $V(x)=1/2kx2$ where k is the spring constant can be applied to two and three dimensional models where the Hamiltonian operator acts in two dimensions to $px^2 + py^2 + \dfrac{\omega^{2x}x^2}{2} + m\omega y^2/2$ are divided by 2m where m is the mass. Note in the spring constant k it equals the mass times the angular frequency or ω^2. *The eigenvalue for $\Psi n x n y(x,y)$ is* $Enxny = \hbar\omega x\left(nx+\dfrac{1}{2}\right)+\hbar\omega y\left(nn+\dfrac{1}{2}\right)$. *The annihilation and creation operators as a and a*produce a harmonic oscillator as the Hamiltonians* $H = \hbar\omega\left(a^{*\prime}a^*+\dfrac{1}{2}\right)$ *and* $H = \hbar\omega\left(a^*a^{*\prime}\right)-\dfrac{1}{2}$ *where* $a^* =\left(\dfrac{m\omega}{2}\hbar\right)^{\prime}/2\left(X+\dfrac{iP}{m\omega}\right)$ *and* A*'=(mω/2ℏ)^1/2(X+iP/mω) yieding a state of |n>=1/(n!)^1/2(a*')^n|0>

A*'=(m yieding a state of |n>=1/(n!)^1/2(a*')^n|0>

Where |0> is the vacuum state which is always approached but never reached.

The harmonic oscillator in the approaching 0 dimensional space or infinite dimensional space of the quantum bubble can be applied as space-time oscillates from an anti-swirl to a swirl unwinding or uncoiling from infinite curvature of space-time at the quantum bubble to almost complete expansion after Planck Time. Utilizing raising and lowering operators with the eigenfunctions of the harmonic oscillator one can use the annihilation and creation operators along with Hermite polynomials to get the eigenvalues of the Harmonic Oscillator for eigenenergies E n=(n+1/2) $\hbar\omega$ *where n=0,1,2,...The Hermite polynomials are* $\lambda=\dfrac{\left(\dfrac{\hbar}{m\omega}\right)^{\prime}}{2}$ *where* h/mc is the deBroglie wavelength and the velocity is the angular velocity ω *while the other Hermite Polynomial is H n(ℂ)where we have a Hamiltonian across n eigenvalues* to produce n eigenstates The solution to the Schrodinger Equation in this case is a series of eigenfunctions which are Ψ $n(x)$=1/ $\pi\lambda^2)^1/$ 41/$\sqrt{2^{nn!} Hn\left(\dfrac{x}{\lambda}\right)e^{-x^2}/2\lambda^2}$ or $(1/\pi\lambda^2)^1/41/(2^n\ n!)1/2\ H\ n\ (x/\lambda)$ $e^{-x2/2\lambda^2}$.

As antimatter is annihilated in the Big Bang to create Dark Energy and matter is created from energy coupled with the Higgs Field the creation and annihilation operators are both used in the 1,2,3 and n dimensional states with the zero dimensional state or null state being in question. The creation operators exceed the annihilation operators with respect to eigenvalues and eigenstates as matter exceeds anti-matter in the pre Planck Time epoch and the uncoiling of infinite curvature space-time with 7.5×10^{11} strings producing an unwinding rotation with a push outward from the Dark Energy produced by the anti-particle anti-particle repulsion. One would use the lowering and raising operators in the equation $|n> = 1/(n!)^{1/2}$ times $(a8')^n |0>$ where again $|0>$ is the null state which may actually be a near infinite number of quantum dots in a lattice that is too small to be measured and can be curled up or compactified to a zero dimensional state.

CHAPTER FOURTEEN

RELATIVITY

Relativity whether General or Special Relativity is everything being in motion relative to everything else. The observer is moving in many time and space frames up to two π radians or 360 degrees compared to what's being observed which is also moving in space and time frames throughout 2π *radians*. Nothing is stationary at any temperature above 0 degrees kelvin which is asymptotic and never can be reached. As a result there is no absolute resting mass. Inertial mass is relative to velocity in a moving frame relative to other moving frames. As the velocity approaching "c" the speed of light boundary inertial mass increases geometrically toward infinity which is another asymptotic function as based on the Lorenzian Transformations mass=mass $0/(1-V^2/C^2)^{1/2}$ SO $1-V^2/C^2$ APPROACHES ZERO CAUSING INFINITE INERTIAL MASS AS THE SPEED OF LIGHT BOUNDARY IS APPROACHED. In addition the Lorenzian Transformation shows time dilation to infinite length or in essence time stops as time=time $0/(1-v^2/c^2)^{1/2}$ and length shrinks towards zero via length=length$0(1-v^2/c^2)$. In terms of moving frames the observer notes accelerated time around him but to the observed time appears to be infinitely slow as in The Event Horizon of a black hole.

In Special Relativity the speed of light is a constant in a vacuum and the laws of physics are the same for every observer in the same relative motion. Also as light is bent by gravity photons act as matter with a non-resting mass, In General Relativity space contracts as time dilates approaching areas of extreme gravity such as a BLACK HOLE. Space and time are incorporated into space-time and are curved by mass the effect being called gravity. Flat space-time would relate to no curvature and therefore no mass doing the curving while infinite curvature related to a point relates to almost infinite mass with infinite gravity and constricted, curved space-time being small as it the event horizon of a black hole or the quantum bubble pre-Planck Time,

Utlilizing the Lorenzian Transformations and the anti-symmetric characteristic of the Einstein Tensor, Einstein's Law of Relativistic Gravity is formed as G a b=R a b-1/2R g a b=8 πT a b *where G a b is the Einstein Tensor R a b is the Ricci Tensor and R g ab is* the space-time curvature variant of the metric g a b. T a b is the stress energy tensor of the metric g a b. Basically inertia-gravity equals 0 when dealing with the push resistance of inertia and the pull resistance of inertia as described by R a b. Again this is based on 8 πT ab=R a b-$\frac{1}{2}$ R g ab=G a b. THE SPACE-TIME CURVATURE METRIC g ab acts in an anti-symmetric action on the space-time curvature variant R g ab such that G ab =R ab +1/2 R g ab in an equal magnitude but opposite direction to R ab -1/2 R g ab so the effect of gravity on inertia is anti-symmetric as covariant and contra-variant tensors are equal in magnitude acting on curving space-time as +1/2 or -1/2R g ab and these metrics act in an additive and subtracting method on the Ricci Tensor R ab representing inertial resistance to push and pull Resistance to .pull with -+1/2 R g ab and resistance to push with -1/2 R g ab where again they are abelian and anti symmetric (canceling in direction) causing the Stress Energy Tensor to be net 0 or the Einstein Tensor. Poisson's Equation which is the derivative operator of a dual vector field $\nabla^2=4\pi\rho$ *where ρ is the energy density of matter the stress energy tensor of matterT ab can be written* as T ab v^av^b$\longleftrightarrow$$\rho$ and R cbd^a v^a v^b$\longleftrightarrow$$\partial$b$\partial^a\Phi$.R cad^a v^c v^d=$4\pi T$ cdv^c v^d *utilizing Poisson's Equation.giving R c d=$4\pi T$ cd THE STRESS ENERGY TENSOR SATISFIES* ∇^(C T) cd=0 and the contracted Bianchi Identity gives $\nabla c(R\ cd-\frac{1}{2}\ g\ cdR)=0$ *The equality of R cd and $4\pi T$ cd would imply $\nabla R=0$ so R and then t=Ta^a is a constant throughout thr universe. G ab=R ab-$\frac{1}{2}$ R g ab=$8\pi Tab$ no longer has a conflict between the Bianchi Identitiy and Law of Conservation of Energy*

The Bianchi Identity implies local conservation of energy. Taking the scalar trace of G ab =R ab-1/2R g ab is R=-8 πT *or the solution wiith the stress energy tensor and applied as R ab=$8\pi i$ mplies Einstein's Equation is R ab=$8\pi(T\ ab-\frac{1}{2}$ g ab T). Also $8\pi T$ is double the Poisson equation of $4\pi\rho$ where the energy density of matter becomes stress energy* stress energy where ρ *acts in an antisymmteric matter where the covariant and contravariant tensors of ρ double in magnitude but are opposite in direction.* magnitude but opposite in direction causing the stress energy tensor T ab The mass as inertial mass increases as the speed of light boundary is approached and never reached. The speed of light constant acts as a boundary which is never reached. As any action must have an equal but opposite reaction space-time itself may resist matter from breaching this "boundary" and may be creating a weakness in space-time wherever anything traveling at just below "c" is resisted by constricting space-time from extreme gravity and inertia. The speed of light constant of just under 3x10^8 meters second is in a vacuum and the speed of light boundary has space-time pinch off toward zero(which would approach a space less vacuum)which is where "c" would be at 2.99x10^8 meters/second or just under 3x10^8 meters/sec. So as a consequence, "c" is just under 3x10^8 meters/second at the cusp of a cone shape configuration of space-time at the central nexus of the cone at "c" which is why there is approaching infinite push at that boundary caused by space-time traveling at over "c" by the value of π. *Experiments done by Dr. Lene Hau* 21 *have slowed photons down to as little as* 38 *miles per hour ina confluent stream of lasers* per hour with a confluent stream of lasers at temperatures which approach 0 degrees kelvin. In a non-vacuum medium although it's difficult to measure photons it may be possible that light can be slowed. For example in a Black Hole at the center of the Milky Way where photons are absorbed by it's extreme gravity it is possible that electromagnetic radiation is slowed down considerably from approaching "c" to considerably slower velocities. Under those extreme pressures there may be phase transitions within each black hole and it was Steven Hawking who mathematically calculated 252 different states of matter inside a black hole. While extremely difficult to prove can electromagnetic radiation be in other states of matter under the extreme pressure of a black hole. Can the Boltzman Equation be used to determine if photons which have a positive non-resting mass and are bent or affected by gravity matter? Or as an alternate explanation as photons travel along space as space-time progressively curves around areas of extreme mass by gravity the extreme curvature of space-time around stars and black holes may cause the photons to curve down to an infinite curvature point with space-time at the Event Horizon. If they(photons)are matter they may occupy a liquid or solid state under the extreme pressure of each black hole, and the velocity of each photon may indeed be slower than $3x10^8$ meters/second as the medium of black hole may slow photons down. Note:If photons have mass and show properties of matter (just as matter show properties of electromagnetic radiation with the deBroglie Equation)then perhaps they can be used in the definition of a very thin PERFECT GAS STATE JUST THE WAY SPACE-TIME was applied as a perfect fluid with the PERFECT FLUID EQUATION.

General Relativity can be shown in terms of reference frames like object A traveling along the x axis at a velocity of 30 kph with reference to observer B who is moving along the x axis at 20 kph. If object A and observer B are traveling exactly along the same path on the x axis the perceived velocity of object A would be 10 kilometers/hour as object A would stay 10 kph ahead of observer B. To observer C who is relatively stationary the measured velocity of object A is 30 kph.

The angle of observation is also important in the measurement such that an observer traveling at 10cm/hour that is 30 degrees away from the object observed traveling at 30 kph the observer measures 30kph(+ or – 10cm/hr) cos θ *where* $\theta = \dfrac{\pi}{6}$ *radians or where* $\theta = 30$ *degrees. The same can be applied throughout the entire 2π radian observational angle for object A*

Object A by the observer whose measured velocity would be added or subtract from the objects velocity depending on whether the observer is moving toward or away the measured source and to move the angle of trajectory of the observer inline with the measured object the cosine(adjacient /hypotenuse)of the angle differential will give the component of the measured velocity. These premises can be brought out ad infinitum measuring moving fluids in curved space by observers that are also moving in curved space with devices that are subject to measurement error due to Heisienberg's Uncertainty Principle.

Unit tangent vectors have to be used along the measured path of osculating planes with reference to the object A when the observer is not on an osculating plane that is perfectly parallel to the osculating plane that is being measured. The tangent vectors would "touch" the curves of the plane that is in motion with it's component vectors which are perpendicular(N for Normal). This subject was discussed previously in the section on osculating planes and their intersections. The equation G ab =R ab -1/2R g ab=8 πT *is a refinement where the Bianchi Identity implies local energy conservation eliminating a problem with the* Eliminating a problem of the Bianchi Identity and energy conservation.

In the equation of everything from the standpoint of General Relativity show curved Lorenzian or Riemannian space-time $\mathbb{R}$ a b c d as following Bianchi's Identity and is antisymmetric as covariant tensors R ab and contravariant tensors $R^{\wedge}cd$ are equal but opposite in direction and the Bianchi identity has the permutations of the 256 components cancel out to zero, while $\mathbb{R}$ abcd=-$\mathbb{R}$ abcd= $\mathbb{R}$ dcba. The equation of flat Mintkowski space-time has R a b c as =$Ra^{\wedge}bc$=$Rab^{\wedge}c$=$R^{\wedge}a^{\wedge}b^{\wedge}c$ all equal in magnitude but opposite in direction. These are acted upon by the so=pace-time curvature variant R g ab which curves with reciprocal curvature and normal curvature depending on the Ricci Tensor representing inertia R ab. Also R a b c=-R a b c =R cba holding the Bianchi Identity and the anti symmteric quality as space-time is directly prooportional to space with space-time=space-time and curved space-time=flst space-time + or – R g ab which is the space-time curvature variant all caused by the inertial mass or Ricci Tensor which is doing the curving via the metric g ab on the curvature variant R g ab. Space-time is inversely proportional to mass as space constricts around areas of high gravity and time dilates to infinity or stops so as areas of high gravity have a huge inertial mass that inertial mass is inversely proportional to space-time in the Relativistic form of the Equation of Everything. Multiplying by $1/c^{\wedge}2$ causes that inertial mass to convert to energy via $e=mc^{\wedge}2$ and can be in the case of The Big Bang can show it to be T $10^{\wedge}19$ GEV or the Grand Unifcation Energy. The total of Riemann space-time nets zero. The scalar trace of R a b c d=R a b c-1/2R g ab /R ab(1/c^2) is R=R-1/2gravity divided by inertia where R =R and -1/2 gravity/inertia causes the curving of space-time which is on the left side of the equation as curved by matter *on the right side.* space-time so spacetime=spacetime; Here gravity and inertia' The energy density of matter ρ *can sub* stutute for the mc^2 in the denominator where the m is inertial mass so the scalar trace can be R=R-1/2R (curvature variant caused by the mass in the energy density of matter)/ ρ *where curved space-time=* curving space-time(by gravity)/ρ(energy density of matter)or the scalar trace is R=-R/ ρ *where* ρ=Rab(c²). *Of course the scalar trace of* $Rabcd = Rabc - \frac{1}{2}Rg\frac{ab}{R}ab$ *is* $R = -\frac{R}{\rho} or -1\frac{R}{\rho}.$ *where R is space-time(4 space) and (3 space)-1/2 gravity divided by the energy density of matter or R ab c^2 from the Ricci Tensor.*

Stress Energy T relates to the metric of gravity in the form g ab which results from the curving of space-time through the curvature variant R g ab from the inertial mass R ab and the cause of the curvature which is inertial mass. Gravity is the effect of the curvature and inertia is the cause of the curvature and this results in the stress energy tensor of matter T ab. Poisson's Equation which is ∇(*dual vector field ab, ba)=4$\pi\rho$ where the energy density of the inertial mass R ab which is is R ab(c²) from E=mc² is multiplied by by 4π normalizing the equation as* $-\frac{1}{2}Rg\frac{ab}{R}ab(c^{-2})$*where R ab (c²)is 4$\pi\rho$ resulting from from 2π radians for each of the vector fieldthr*

Fields where ρ=R ab(c²) *so in a dual vector field R a b c d=* $Rabc - \frac{1}{2}Rg\frac{ab}{2\pi\rho}$ *and* $Rdcba = Rcba + \frac{1}{2}Rg\frac{ab}{2\pi\rho}$ *for the mirror image according to Bianchi's Equality for space-time. In these cases the scalar trace is* $R = -\dfrac{R}{2\pi\rho}$ *as direction of the fourth degree tensors doesn't matter in direction only onl magnitude.*

In the case of a dual vector field R a b and R b a the covariant tensors are important and the contravariant tensors R^c d and R^d c cancel out as zero(0). So space-time R ab^cd+Rba^cd=R ab^c+R ba^c-1/2R g ab/4$\pi\rho$ where the contravar iant tensor fields of c cancel out in equal but opposite directions leaving R ab and R ba where R ab=-R ba and R ba=-R ab for space-time which is curved by the metric g ab in the Ricci Tensor R ab on space-time R ab via the curvature variant R g ab. Therefore the final form of the Relativistic Equation representing space-time=space/mass(c^2)is Rab^cd+R ba^cd or R ab^cd is – Rba^cd substituting for Rba^cd yielding Rba^cd-Rba^cd+Rba^cd or –R ab^cd+R ab^cd equaling zero(0) for the Riemann Metric of Space-time following the Bianchi Identity. this means that Rabcd=0=R a b c-1/2 R g ab/4 $\pi\rho$ *where* 4$\pi\rho$ *is R ab(c²)incorporating the Grand Unification Energy or* 10^{19GEV}.

THIS IS THE RELATIVISTIC FORM OF SPACETIME=space/ mass(c^2) or THE EQUATION OF EVERYTHING TIED IN WITH A CIRCLE WHERE THE CIRCUNFERENCE IS 2π*RADIANS AND THE ENERGY DENSITY OF MATTER IS THEE DIAMETER R a b(c²)*

So 4 $\pi\rho$ *describes the diameter of the circle* 2$\pi\rho$ *describes the radius and* 2π *radians describes*

The circumference or space-time. As the circumference /radius $= 2\pi \frac{radians}{2\pi} \rho$ *the 2π's cancel out leaving ρ in the denominator for the Riemann Forces $e=mc^2$ and 1 in the numerator for space-time with all curvatures incorporated by all mass ρ. This equation $s=r\theta$ is another form of The Equation of Everything* which incorporates M Theory and type II a String Theory compactified to a circles the circumference is all inclusive space-time with with all curvature from all mass and the radius is ρ *the energy density of matter then the diameter is 2ρ so* $\pi = \frac{1}{2\rho}$ *or* $2\pi\rho=1$. The circumference 1 is all space-time inclusive with all curvature and from ρ *the energy density of matter and $2\pi\rho=$circumference or space-time$=1$. Curved space-time$=2\pi\rho$ or 2π(energy density of matter)$=2\pi\rho=$total space-time where the curvature variant is θ in $s=r\theta$ and $r=\rho$.*

Set space-time=space-time and make s=space-time=$\mathbb{R}^4$. *$Space-time = 2\pi\rho = r\theta.\rho = \frac{r\theta}{2\pi}$. The Riemann Forces suggest* $\mathbb{R}^{fR}g\frac{ab}{2\pi}orR(f)Rg\frac{ab}{2\pi}.Rab + Rg\frac{ab}{2\pi} = stress$ *energy. This follows* $8\pi Tab = \frac{8\pi(Rab+Rgab)}{2\pi} = 4(Rab+Rgab)$. $\nabla(a,b)=4\pi\rho$ *via Poisson's so* $\rho = \mathbb{R}fRg\frac{ab}{2\pi}(c^2) = RabRg\frac{ab}{2\pi}$ *and stress energy is $4(Rab+R\ g\ ab).\rho$* $4(Rab+Rgab).\rho = \mathbb{R}\frac{f(Rgab)}{2\pi}$ *where* $Rf = Rab(C^2) = r.So\rho = \mathbb{R}(f)c^2Rg\frac{ab}{2\pi}and\rho = Rab(c^2)$
So as a result The stress energy tensor is a dual vector field which is expanded to 4 space

Space via $\mathbb{R}4$ *with regard to the metric g a b has the energy density of matter $\rho=2(R\ ab+R\ g\ ab)=T\ a\ b$.* Exapnding this to a tensor of the fourth degree with the Bianchi Identity applying for $\mathbb{R}4$ *with regard to (a,b) as the metric g one gets $8\pi T\ ab$ as Poisson's Equation states*

A dual vector field $\nabla(a,b)=4\pi\rho$ *and in* 4 *space two of the fields are equal but opposite in a* 4x4 *matrix or* 8π *T ab*. This 4x4 gives 16 permutations and these permutations have three other 4x4 matricies based on g ba,g^a b,g^,ab where the contravariant tensors are equal but opposite to the covariant tensors. This is limited to the metric or(greek measured) g ab whose curvature variant R g ab= θ *in the expression* $s=r\theta$ *or* $\mathbb{R}\infty=\Sigma$ *Rn(R g ab)where n* 0 *to*∞ *eigenstates(approached asymptotically)over n dimensions.*

Again this is based on the almost infinite dimension of intersections of a subdivided 1/3600 degree of arc in expansion and rotation where the lines are non parallel due to θ *or the space-time curvature variant R g ab. As*$\theta\rightarrow0$ *space-time approaches flattness as in the vacuum state or eigenstate* $\rightarrow0$ *in which case* $2\pi(\mathbb{R}(f)\rightarrow0$ *circumference of the space-time circle makingit a linewhich is infinitely short(length*$\rightarrow0$)which is *the state of the Lorenzian Transformation as v*$\rightarrow$*c which is the speed of light boundary PROVING THAT SPACE-TIME CONTRACTS TO ZERO* (0) *OR A POINT OF INFINITE CURVATURE AT "c"*

CHAPTER FIFTEEN

GRAND UNIFICATION ENERGY AND YANG MILLS TERMINOLOGY

GAUGE SYMMETRY GROUPS in Yang Mills Terminology describes the eigenvalues of energy. In terms of gauge symmetry groups the Grand Unification Energy which is 10^19 Giga electronvolts describes the energy involved in all subgroups of eigenstates U(1) x SU(2)x SU(3)x SU(4)x SU(N) x Osp(1/4) all in Non-Abelian Lie Groups. Described as the GUT, the energy of "The Big Bang" is described in terms of the strong force, weak force, electromagnetism, gravity and other weak perturbations involved in quantum fluctuations. U(1) describes the force of electromagnetism which is a strong force while SU(3)describes the strong force associated with nuclear fission and quarks. The electroweak force associated with nuclear decay is represented by U(1)x SU(2) in terms of gauge groups and gravity is considered in the SU(2)gauge symmetry group. In this instance gravity is considered a weak force although by definition it is still an effect and isn't clearly treated as such in gauge symmetry. In the unified forces of the GUT SU(3) is considered a super-strong force and may be both attractive and repulsive depending on whether dealing with matter or antimatter as gluons and quarks should have anti-gluons and anti-quarks where the curves and reciprocal curves of space-time pull or push matter in terms of ρ *the energy density of matter.* 10^{19} *GEV should approximately equal* $6.75x10^{34}$ *erg*

Expansive force of the "Big Bang" with the uncoiling of the rotational or centripedal force of the quantum bubble at Planck Time 10-43 seconds. $\mathbb{R}$*considered ALL FORCES AS* (*Riemann Forces*) *are considered synonymous with R* (*Riemann Curved Space*) *as Riemann*

ved space-time)as Riemann a German Physicist who died of Tuberculosis at age 40 believed that all forces are twists and turns in the matrix of curved Riemann space rather than just gravity as the curvature of space-time by mass as described by Einstein. Therefore $\mathbb{R}n=\mathbb{R}n'$ *where n is dimensional Riemann space-time and n' are the sum total of all Riemann Forces which should equal the Grand Unification*

Energy in gauge symmetry groups of Yang Mills Terminology.

In terms of metric tensors $\mathbb{R}4=\mathbb{R}(F)=U(1)xSU(2)xSU(3)xSU(N)$ $xSO(32)xOsp(\frac{1}{4})=$ *GUT where Rabcd=-Rdcba and R* $ab^{cd}=$ *-Rcd*^{ab} *and for a metric g ab R g ab in* $\mathbb{R}4$ *reveals cd=-dc and ab=-ba with regard to antisymmetric properties in the abelian groups. In these cases where* there are 16x16=256 permutations of Riemann in the case of g a b the component of g cd must be taken into account in Riemann 4 space $\mathbb{R}4$ *and from Newton*$^{\wedge'}$ *sThird Law the reaction which is equal but opposite to the action of the* metric g ab is S=-1/2k^2(-g)^1/2R as the action and the reaction is S'=-1/2k^2(--g)^1/2R or S'=-1/2k^2(g)^1/2R where R has reciprocal curvature of space-time to the curvature invariant of R, Notice the coupling constant of gravity k stays the same. As a consequence of this R a b cd becomes a 4x4 matrix with Rab^cd,R cd^ab,Rad^cb,Rabc^d,Rcba^d,Rdca^d etc. and with g ab having an action in 4 space rather than 2 space as in flat strings such as the orbifold many of these components of this matrix will zero out such that R ab^cd=R ab^00 where cd=-dc=0. This is what makes the extension of Riemann Forces with parallel transport onto Riemann Surfaces so involved and intricate. As far as Yang Mills Gauge Symmetry the involvement of tensors of the 4th degree with covariant and contra variant components is simplified into the internal symmetry group for Super symmetry such as SU(3) and the Poincare and Lorenz Group. These supergroups are also considered Lie Groups as in Lie Algebra and with real parameters such as the super groups Osp(N/M)and SU(N/M). The SU(2)x U(1) where electromagnetism is incorporated in U(1)XSU(2) with quarks associated with spin up and down iso-spins for different flavors are associated with the Strong Force of SU(3) in non-abelian gauge subgroups. The SO(32)is a description of String Theory which has mutual duality with type I, type II, type II a and the Heterotic 8x8 and is described as a whole by M Theory. It was listed as a gauge symmetry group or non Abelian Lie Group in terms of an energy subgroup of the Grand Unification Theory just s super symmetry or super gravity would be listed as subgroups Osp(N/M) or SU(N/M).

There is a mass gap problem in Yang Mills Theory associated with $\mathbb{R}$4. *The mass gap Δ>0 in a compact simple gauge group G where a* *non-trivial Yang Mills Theory exists. The mass gap*Δ *is the mass of the least massive particle predicted* by the theory. In ℝ 4 *assuming no "shadow" dimensnions which would be odd dimensions* 1 *an*3 *upward containing* dark matter as discussed in a later chapter one would have to consider PHOTONS AS POSSIBLY FILLING THE MASS GAP. Photons bend or curve space-time and are absorbed by Black Holes therefore having a mass when it motion albeit miniscule and no rest mass(photons have not been stopped even at temperatures near 0 degrees kelvin). So if photons can be defined as matter the entire gauge group U(1)involving electromagnetic radiation would be rife with photons with a mass under 10-27 kg/photon per experiment. Remember light is bent by gravity. As photons travel at just under "c" they should follow the Lorenzian Transformation showing inertial mass increasing geometrically as "c" is approached which means the mass of a continuous photon stream in constricting space-time would buffet against the inertial mass of space-time above "c" causing the photon mass to increase dramatically. Since there is no way to experimentally determine the mass of each photon stream in constricting space and dilated time we can't determine if this is true of photons in this state although the mass measurement of photons in the laboratory was done at extremely cold temperatures with the confluence of lasers which slow photons down to at times as little as 36 miles/hour. Based on this it would be easier to prove photons account for the mass gap than dark matter in shadow dimensions. If the intelligent observer were travel at a velocity approaching "c" just above the speed of light the intelligent observer would be able to measure considerable mass for photons which cannot be measured easily when the observer is relatively stationary.

Photons can be considered as possibly the next energy level past the ground state(vacuum state) in the mass gap. The gauge symmetry group $1(J^{\wedge}PC)=0,1(1--)$follows the "massless" photon whose true non-resting mass $<10^{\wedge}-18$electron volts/$c^{\wedge}2$ and stable over lifetime of the photon. The photonic spin is $+1$ with a charge of $<10^{\wedge}-35$ e. The Parity is -1 and C parity is also -1 with no resting frame as photons have never been shown to stop completely even in experiments at Harvard. As a result the CPT Theorem holds as the time element would be the same and the direction of motion of a photon is $\pm\hbar$ *Or Planck$^{\wedge'}$ sConstant) and the direction is a double helix much like DNA. In terms of space-time curvature this miniscule mass burrows through space-time as a double helix but space-time is contracting as it travels at or near "c" contracting around the double helical photonic stream*

Stream which then contracts to a spiral or cone configuration at "c" just as at the event horizon of a Black Hole. The equation for photonic mass is $E^{\wedge}2=p^{\wedge}2c^{\wedge}2+m^{\wedge}2c^{\wedge}4$ where $m^{\wedge}2c^{\wedge}4$ do not vanish and where the momentum $p=mv$ and the mass is not negative but miniscule. The equation$< \phi(0,t)\phi(0,0)becomes<\phi(o,1(1--))\phi(0,0)>aga$ *in where the gauge symmtery is* $1\left(J^{pc}\right)=0,1(1—)$ *and this approximately equals* $\Sigma nA \; n \; e^{-\Delta nt}$ *in the mass gap.*

It is difficult to negociate string theory with Yang Mills however the S0(32)string theory incorporates gauge symmetry and therefore a non-Abelian gauge group can be incorporated into Yang Mills Thoery and Quantum Field Theory. Therefore the mass of a string must also be considered in the mass gap. The mass of an open string is based on the following operator:$m^2=2$

Vector product a-n^i an^i *where a^i s the Regge Slope which is $2\pi T=1/$ $a^{\wedge\prime}$ or the reciprocal of 2 π times the tension of the string in two dimensions. a' is Regge Slope not $a^{\wedge}i$*

The mass of a string is the Tension (T)x length of a string which is $10^{\wedge}$-33 cm and the mass is an eigenvalue with the operator acting on string energy states or eigenstates. *The expression$a$-n^{iani} are integer valued excitation levels of energy of the quantum harmonic oscillator*

Which is associated with what's called the Fourier Mode of coordinates x^2 along an open string in flat space acting as a standing wave. The open strings are Bosonic strings as in the expression of gravity's effect of the mass on flat space. Recall that the spin 2 vector boson is associated with space-time curvature. For closed strings the mass m^2 sums over chiral oscillators which move in opposite directions. With D= the D dimensional state and I being the initial dimensional state the term SO(32)can be incorporated into Yang Mills Terminology and the mass gap Δ*for strings would be*<$\phi(m^2,t)$ $\phi(0,0)$>= SO(32)for closed strings acting in a Bosonic and Ferminic Field where the Fermion is the vacuum state and the Bosonic field reflects the curvature of flat space-time as Hilbert Space into the Orbifold where the string mass at a low energy level curves Hilbert Space. The one dimensional state I mimicks the infinite dimensional space compactified at under Planck Length which may or may not be a boundary for space-time not to break up into component quantum dots. If the eigenvalue for energy for state i →i+1 *reflects the smallest mass then the mass gapΔ would be based on the open or closed string* rather than the photon as having the smallest mass.

Gauge Symmtery involves elementary particles where there are patterns in these particles. It is analogous to a confluence of mirrors where the sequence of reflections form a pattern where spin acts similar to mirrors and the mirrors from isospins + or - has appearance of symmetry pattern. Symmetry relationships for multiple subatomic particles form patterns which are called symmetry groups. With regard to quarks the different flavors and reflections form the SU(3) gauge symmetry group. This particular group represents symmetries of the groups of gluons and strange quarks, up quarks, down quarks etc. and all gauge symmetry groups handle the subatomic particle components of that group and not other groups which is why the groups are non-Abelian. The omega minus particles

Members with the up quarks, down quarks and strange quarks plus the Ω-*particle*. Neurons and protons make an octet of heavy fermions(fermionic fields also indicate the vacuum state). Eight different varieties of mesons form another octet. In the SU(3) subgroup for the strong force comprise of symmetry patterns for 3,8 and 10 particles. Quarks have interactions through hadrons and sometimes groupings of quarks comprise symmetry patterns through the iso spin reflective patterns causing the symmetry. Ω-*particle may contribute to the mass gap Δ>0 previously mentioned but not clearly identified. If one point in space is identical to another point in space*

Point with local symmetry forms global symmetry and local symmetry involves choices regarding subatomic particles forming fields. The members of a local symmetry group follows the same rule regarding their choice such as in the U(1) subgroup involving electromagnetism all members of that subgroup form electromagnetic fields and follow the same rules as for all electromagnetic fields. Of course these are called gauge fields. Please note that photons in electromagnetic fields act as though they have a miniscule non-resting mass and appear to be incorporated into then U(1) subgroup of gauge symmetry, which is why the mass gap is less likely to be from photons and more likely from strings.

CHAPTER SIXTEEN

PHYSICAL COSMOLOGY AND THE HUBBLE CONSTANT

When one considers "The Big Bang" one must consider the physics of the cosmos which includes the rotation and expansion of space-time with the quantum bubble containing approximately 7.5×10^{11} strings and a force of 6. 75×10^{34} erg in the first second after Planck Time (10-43 seconds)although the total force was considerably larger and space-time expanded at a speed from "c" the speed of light boundary to πc *with matter being propelled outwardly with a trajectory of πradians due to the mutually* repulsive force of antimatter anti-matter repulsion in a bubble that was 10-33cm or Planck Length times 7.5×10^{11} strings. There was a maximum moment of rotation at just before Planck Time and the rotation reduced and uncoiled geometrically while the expansion increased geometrically either through an orb blast or cosmic inflation. There is an approximate mix of 50:50 between matter and antimatter in the quantum bubble with the matter attracting other matter via gravity, and antimatter repelling antimatter with anti-gravity and the anti-gravitational effects of the "Big Bang "causing Dark Energy which is the force pushing galaxies away from each other far in excess of the weak force of gravity in the remaining matter which will attract.

The Friedman type II open flat expanding universe has in general been acknowledged as at least asymptotically similar to the universe we live in. In this scenario space-time has no curvature from mass and anything shot out in a completely straight line will continue on that line if there are no significant unbalanced forces until resistance slows the bullet down to stop it eventually and if outside the pull of earth's gravity the bullet will never return to the starting place in Euclidian Space or flat Mintkowski Space-time with a force acting on it to return it to earth. Albert Einstein had an alternative hypothesis for the universe called "Einstein's Closed Curved Universe" where boundaries exist and something shot out will eventually return at or near the region in space-time where it began without unbalanced forces just based on curved space. The question is this ;why does space appear to be completely flat when in reality it is curved as space-time? In 1921 Albert Einstein predicted a positional change in celestial body near the sun(solaris)which can be detected during a total solar eclipse which had to be repeatable to avoid measurement error and would prove space-time curvature rather than flat Mintkowski space as existing between the sun and the other celestial body. The experiment was successful and space-time curvature was accepted by the scientific community as being a proven fact. The amount of curvature of space-time can be calculated if Dr. Romodo Cortez de Pauai's figures

For the mass of the universe are correct at 10^{54} kilograms. Based on this ano Einstein's Equation of Relativistic Gravity one can calculate 8pi T=G ab=0=10^{54}-1/2R g ab where R g ab=10^{27} where R ab=inertial mass and T ab is the stress energy tensor between inertia and gravity. Applying this to the equation R a b c d=R ab c-1/2R g ab/R ab which is the Relativistic Form of the Equation of Everything R a b c d=R a b c-0.5x10^{27} kg/10^{54}kg or R a b c-5x10^{26} kg/10^{54} kg or R a b c/10^{54} kg-0.5x10-28 kg =R a b c/10^{54}-5x10-29 and since Riemannian Space-time in Riemann tensor in 4 space =0 then in this case where Einstein's Tensor for Relativistic Gravity=0 the space-time curvature variant R g ab on R a b c as flat space-time is approximately zero but not zero rendering it asymptotically flat. R g ab is actually 5x10-29 meters in the MKS system based on a 54 kg universe which is extremely close to 0 flatness or asymptotic flatness but it isn't flat in actuality. The space-time curvature metric is more pronounced at the Event Horizon of Black holes where the pull of gravity from the collapsed mass funnels space-time down toward a point of infinite curvature in that region. The regions of space containing black holes vs space not containing black holes are such that the black hole regions are very scarce as are stars and even galaxies compared to the void of expanding and slightly rotating space-time such that the curvature metric is so small that it would escape detection except under the most extreme conditions but it is there.

Please note also that the calculations with the mass of the universe being 10^{54} and using Einstein's E=mc^2 for the Big Bang the mass of the universe is 3.3x10^34 kg too high based on Relativistic Gravity calculations. This mass is unaccounted for and has in the past been attributed to Dark Matter which can only be detected indirectly with gravitational measurements of revolving bodies and through gravity cameras. It has been postulated that dark matter is partially comprised of baryonic particles and neutrinos however the exact composition is unknown.

CHAPTER SIXTEEN CONTINUED

THE EFFECT OF DARK MATTER AND DARK ENERGY IS LIKE LIGHTING A MATCH WHERE THE PHOSPHORS ARE THE ANTIMATTER AND MATTER, STRIKING THE MATCH IS THE BIG BANG, AND DARK MATTER IS THE CHARRED RESIDUE OF THE MATCH WHILE THE FLAME IS DARK ENERGY(FROM ANTIMATTER)and HEAT(FROM MATTER PLUS THE EXPLOSION FROM MATTER ANTIMATTER ANNIHILATION)WITH ANTIGRAVITY FROM ANTIMATTER ANTIMATTER REPULSION. AS DARK MATTER IS VERY LOW ENERGY IT's COMPONENT PARTICLES MUST BE VERY LOW ENERGY AND THESE PARTICLES MAY BE THE NEXT ENERGY LEVEL ABOVE THE VACUUM STATE IN THE MASS GAP OF THE YANG MILLS THEORY. THE Ω-*PARTICLE MAY BE AN ELEMENTARY PARTICLE IN DARK MATTER BUT IT MUST HAVE CHARACTERISTICS IN COMMON WITH* CHARACTERISTICS IN COMMON WITH QUARKS IN THE SU(3)SUBGROUP. QUARKS AND GLUONS KEEP NUCLEII TOGETHER IN ATOMS AS DARK MATTER IS PURPORTED TO ACT AS COSMIC GLUE LIKE GLUONS. The mass gap in Yang Mills must be as pervasive as the missing mass in the universe for this to hold. The mass gap is between the vacuum state and the next lowest energy level with the smallest mass possible for a particle. If dark matter has so many homogeneous particles that they cannot be directly detected, these particles may be so small that they meld or mesh into space-time or in the case of Yang Mills Euclidian space.

Based on this it's possible that dark matter would most likely account for the mass gap in Yang Mills Theory although it would occur in multiple gauge symmetry groups but least in U(1)as they have little in common with electromagnetic radiation. With regard to electroweak forces of nuclear decay and gravity the dark matter particles may form an integral part of the U(1)x S(2) and S(2)gauge symmetry groups. Being non-abelian though they must share reflective symmetry patterns with other members of these groups. The temperature of space is 2.74 degrees kelvin. Dark Energy is so pervasive in space that unlike other energy it must be COLD. Scientifc measurement can not verify high energy with Dark Energy. Dark Matter is extremely low energy and possibly extremely compact small mass particles which may stick or adhere to other particles. In terms of gauge symmetry the grouping would be U(1)xSU(2)xSU(3)x...SU(n) as it is likely associated with electroweak forces, gravity and The Strong Force. These microparticles may be the *Ω-particle in SU(3)and may relate to all matter except for photons. Also photons and electromagnetic radiation* electromagnetic radiation can be of very high or lower energy levels and the energy level of a microparticle of dark matter may be below that of a cold photon. The mass of this microparticle would be the total mass of dark matter or 3.3×10^{24} kg/number of particles per volume of the universe where the volume of the universe is derived from *ρc(critical density=mass of the $\frac{universe}{volume}$ of the universe. So ρc=mass/volume* and from Poisson's Equation $4\pi\rho = \nabla\left(\frac{mass}{volume}\right)$ *where mass and volume are the dual vector field. Mass can=the Ricci Tensor R ab and volume is* $\mathbb{R}4$ *space. Therefore* $4\pi\rho = \frac{Rab}{\mathbb{R}} = R\frac{ab}{Rabc} - Rg\frac{ab}{R}ab$ *ab or the vector product of* $R\frac{ab}{Rabc} - \frac{1}{2R}Rgab + \frac{1}{2}Rgab$ where R g ab is antigravity and –R g ab is gravity suggesting reciprocal curvature and curvature of spacetime ℝ. *R g ab>-R g ab as antigravity causes H0 and theΩ of the universe. Therefore $4\pi\rho=R$ ab(as a vector product of R ab.R ab)* $-0+\frac{1}{2}Rgab$ *as antigravity exceeds gravity due to H(Hubble expansion at time* $\frac{t}{H0}$ Hubble expansion at time 0. So it follows that

$$4\pi\rho = Rab.Rab\frac{1}{2}Rg\frac{ab}{R} abcorRab.Rab +$$

$$\frac{1}{2}Rgab \div Rabcor\rho = Rab.Rab + \frac{1}{2}Rg\frac{ab}{4\pi}Rabcor\rho = Rab' +$$

$$\frac{1}{2}Rgab \div 4\pi RabcwhereeRab' = Rab.Rab.$$

R ab=3.3x10^24 kg for Dark Matter Rab'=10 ^54 kg for the universe *ρc=critical density of the universe and ρ=density of dark matter.* $\frac{\rho}{\rho c} = Rab \div Rab.Rab = 1 \div Rab$ $1 \div R$ ab which is $\frac{1}{3.3x10^{24}kg}$ $or.33x10^{-24} = \frac{\rho}{\rho c}$ *which approaches 0 density for dark matter yet it has extensive mass which is* in line with scientific observations. The particles would be so small that they would need to be compactified to 0 (curled up) with 0 density, positive mass and positive gravity. As particles or microparticles of dark matter →∞ *they may be below Planck length 10-33 cm in size and very low in energy* causing the mass gap Δ>0 in Yang Mills Theory.

The Hubble expansion is such that $H^2 = \frac{8\pi}{3m}^{2\left(1+\frac{\rho}{2.9}\right)} + \Lambda4 + \frac{E}{a^4 or}$ or $8\pi\left(1+ñ \div \frac{E}{2}\sigma\right) + \Lambda4 + \frac{E}{a^4 w}$ *herem = masss,* ρ *is the energy density of matter, e is energy, a is cross sectional area Hubble expansion parameter and* $H2=\left(a.\frac{}{a}\right)ora' \div a = \frac{8}{3\pi G\rho e}$ *here mass density is* ρ (x,t) $== \rho b(t)\{1+\delta(x,t)\}$ *so* $== \rho b(t)\{1+\delta(x,t)\}$ *so* $\frac{a'}{a} = \frac{8}{3\pi G\rho B}$. *or* 8÷3(πG) (ρB)*where 8/3 is multiplied by πGρB. and the time evolution of the expansion parameter*

I (as a(t) where a is the expansion parameter in the expression a./a or a'/a with a being the area being expanded. $\frac{a(dot)}{a})^2 = 8/3\pi G\rho B + 1/a^2 R^2 +$

$\ddot{E}/3.a(t) = a0[1 - H0(t0-t) - q\frac{0Ho^{2(t0-t)^2}}{2} + \ldots or[1 - H0(t0-t) - q0H0^{2(t0-t)^2} \div 2\, where$

t 0-*t* *is the lookback time and the universe based on redshifts is*

$H0\int_{u}^{} \frac{du}{u} \int_{}^{\infty} dy/y[\Omega y^3 + \Omega(R)y2 + \Omega(\Lambda]1/2.$

$\Omega(R) = \frac{1}{(a0H0R)^2}\, and\Omega \wedge = or\Omega(\Lambda) = \wedge/(3H0)^2$ $=\Lambda/(3H\,0)^2.$

$\rho c = \frac{3H^2}{8\pi G}\, and\dot{U} = \frac{\rho}{\rho c} = 8\pi G\rho/3H^2$

Peebles's 23

As a perfect fluid p $p\Lambda = \frac{\Lambda}{8\pi G}\, andp\Lambda = -p\Lambda and 4\pi G = \Lambda = \frac{1}{R^2 w}$

So $H^2 = 8\frac{\pi}{3m^{2\left(1+\frac{\rho}{\sigma}\right)}} + \Lambda\frac{4(in4space)}{3} + \frac{E}{a^4}$ *has* σ*relate to the pressure gradient*
on the energy density of matter from the

Expansion of space-time based on the Friedmann Equations.
Rewritten $H^2 = 8\frac{\pi}{3m^{2(1+\rho\div2\sigma)}} + \Lambda 4 \div 3 + E \div a^4$

H^2=8(pi)G/3m^2(!+p(rho)/2(sigma)+cosmologic constant in Riemann 4 space/3+energy/a^4

Einstein's Perfect Fluid Equation is R^ij=k(T^ij-1/2g^ijT)+Λg ^ij

T^ij *is stress energy* κ*=8*π*.k=Gaussian curvature in length* ^2. σ *relates to whose effect is to increase dampening experienced by the scalar field*

as it rolls down its potential where $\rho = \frac{1}{2\dot{\phi}^2} + V(\phi)$ *and the potential is V(*φ*)so* σ*is the gradient of the field* $\rho = 0.5\dot{\phi}^2 + V(\phi)$ *is the expression for the energy density of matter with regard to the scalar field* φ

The initation of the Big Bang with a rotational vector in the quantum bubble

An infinitesimal roation through a small angle d θ *about a rotation axis with the direction cosines c1, c2, c3 is described by the orthogonal infinitesimal transformation x' (1+infinitesimal transformationx'=(1+Δ)x* so that Δ *is skew symmetric. With a reference frame with elements a1, a2, a3 which are orthonormal*

mutually perpandicular Δ *is represented by the skew symmetric matrix* $\Delta = \begin{smallmatrix} 0 & -a3 & a2 \\ a3 & 0 & -a1 \\ -a2 & a1 & 0 \end{smallmatrix}$ *for* $\theta = d\theta \to 0$ *such that* $\Delta x = (a \times X)d\theta$ *where a=a1u1+a2u2+a3u3 which is a unit vector in a+rotation axis. For a continuous D3 rotation at time 0 x'(t)=A(t)x where x=constant giving* $\dfrac{dx'(t)}{dt} = \dfrac{dA(t)}{dt}$ *times x=ω(t)x x'(t)=ω(t)x [A(t)x].ω relates to angular velocity and the vector ω(t)is given with reference to fixed and rotating vectors* direcected along the instantaneous axis of rotation (axis x' to x'+dx') and *ω(t)| or |ω(t)is the instantaneous absolute rate of rotation.* It follows that aA(t)/dt= Ω(t)A(t)*with Ω as the skew symmetric operator with reference to*u1, *u2, u3 and u'(t), u2'(t) and u3'(t)*

u3'(t) as elements on which the skew symmetric operator with respect to time works as a 3x3 matrix for *Ω(t)and Ω(avg)(t)so if A-1 power is called* $\dfrac{A(\sim)dA}{dt} = \dot{U}A = A\dot{U}(bar) and \dot{U} = \dfrac{dA}{dt}$ *A(bar)and Ω(bar)=A*$\sim \dfrac{dA}{dt}$ *or* $\dfrac{A^{-1}dA}{dt}$ *the three dimensional Rotation Group-relection group R3± has* proper rotations [det(A)=1] which is a proper rotations group and is a subgroup of R3+- the three dimensional rotation group R3+.R3+- and R3+ don't follow the communitive law.

Space-time has constricted space curved by mass of stringx7.5x10^11 strings from infinite curvature to near flat space-time. $\mathbb{R}4 \leftarrow \mathbb{R}3$ where $\epsilon(t) \leftarrow 0$. $\int_{\infty}^{0} \frac{du}{u} = \ln \infty - \ln 0 = \infty$ curvature for

A point with infinite curvature at the point of maximum curvature

And approaching 0 curvature as space-time approaches flatness. At $\mathbb{R}3$ *the three dimensional rotation group applies with dimension*

N=2j+1 and a 2j+1 by 2j+1 matrix represents the rotation of Cayley-Klein parameters. In this case j=1 for N=3 and the unitary irreducible representations are $\mathbb{R}\frac{1}{2},\mathbb{R}1,\mathbb{R}\frac{3}{2},\mathbb{R}2,\mathbb{R}\frac{5}{2},\mathbb{R}3$ *withCayley-Klein parameters a,b,-b*,a*and Euler Angles*
$\alpha,\beta,\gamma.\mathbb{R}j = xj\left(\alpha,\beta,\gamma\right) = Tr[U^{j}mq\left(\alpha,\beta,\gamma\right) = \sin\left(j+\frac{1}{2}\right)\delta / \sin\delta(.5)$

where $j = \frac{0,1}{2}, \frac{2,1,3}{2},...)$ *where δis angle of rotation.j=.5,1.1.5 and m,q=-j,-j+1,-j+2...j-2,j-1,j.Tr is the trace(spur).of the 2j+1⊗2j+1 matrix*

The above is spherical surface harmonics of degree j. For integral values of j the functions represent a (2j+1)dimensional representation for space $\mathbb{R}(j)$ *with orthonormal functions. Direct products of* $\mathbb{R}3$ *rotation groups describe the*

Composite rotations of dyanamic systems such as the pre-Planck time quantum bubble. The Clebsch-Gordan Equatiom is $\mathbb{R}^{j}\otimes\mathbb{R}^{j'}$
$=\mathbb{R}^{\wedge}(j+j')\oplus\mathbb{R}^{\wedge}(j-1)\otimes\mathbb{R}^{\wedge}(J'-1) \ \mathbb{R}^{\wedge}(j+j')\oplus\mathbb{R}^{\wedge}(j+j'-1)\oplus\mathbb{R}^{\wedge}(|j-j'|)$

APPLY THE ROTATING VECTORS WITH THE SKEW SYMMETRIC OPERATOR Ω *FOR J=1 or three dimensions of Riemann space* with vectors ω 1(*t*), ω2(*t*)ω3(*t*) *where t←0 using* u1, u2, and u3 in a 3 $\otimes$3*matrix*. Applying the mass of a string $2\pi T\Sigma n=1$ *to*$\infty\Sigma i=1$ *to* D-2 α-n^i αn^i where α^l=*Regge slope and* $2\pi T\rightarrow\frac{1}{\alpha(one)}$ *gives the square of the mass of an open string in two dimensions and is the squared*

Mass operator acting on the composite rotations in three dimensions for Riemann space at time time 0 to time t. This value reflects almost infinite curvature of spacetime with 7.5x10^11strings and the mass operator of an open string reflecting the Ricci Tensor for open strings. n is the integer valued excitation level of the quantum harmonic oscillator which for closed strings m^2 has different oscillations with regard to n(2 πT *with a standing wave function and with two chiral components* (*mentioned previously*)

Left and right handed moving oscillators going from the fermionic vaccum state toward the boson state as the mass curves Riemann space and uncurves Riemann space where the curvature is increased by gravity of matter and uncurvature is increased by antigravity of antimatter. The Dark Energy pushes out the quantum bubble from its rotational state to an expansive state in 10^-43 seconds as the infinite curvature of space-time uncoils to almost flatness. The ends of the quantum bubble(polar regions)rotate in opposite directions one metric g a b for matter provides curvature of spacetime and other metric =-g ab for antimatter forming a twisted double torus with the center being the weakest point where "The Big Bang" outward pressure is pushed. R g ab=-R g ba for gravity effect with spacetime curvature metric. For antigravity the antimatter has a curvature variant of R(-g) ab=R (-g)ba=-R g ab. The infinite momentum limit(R)relates as the reciprocal of the Neven-Schwarz 5-brane 1/2NS -NS 2 potentials relating to the fermionic or vacuum state with D-0-Branes. R ab is the mass of all strings and the angular velocity $\omega 0 \rightarrow \omega t$ *gives a centripedal force* $2\pi T \Sigma \omega 0 \rightarrow \varpi \, t \Sigma \, 1 \rightarrow D2\alpha \, \omega 0 \, \alpha \, \omega \, t \rightarrow 0 = m^2$ *operator on* ω *and* r=*radius of quantum bubble is* 10-33cm $(7.5 x 10^{11 strings})$*giving radius* $7.5x10^{10-22}$ *cm with massof string*$(\varpi \, t)^2 \div r$ *makes centripedal force* $7.5x10^{22}$ *erg*$(\varpi \, t)^2 \, m^2$. $R=m\varpi \rightarrow 0$ *as massmof a string* $\rightarrow 0$ *so* $\frac{1}{R} \rightarrow \frac{1}{0} \rightarrow \infty$ *or the infinte momentum limit. D-0-branes go to D-5 branes as uncoiling of spacetime by antimatter occurs.*

CHAPTER SIXTEEN CONTINUED;THE HUBBLE EXPANSION COEFFICENT=The Hubble Expansion Coefficient is the rate of expansion of space-time and the composition of the Universe since the epoch of "The Big Bang"

which was approximately 13.7 billion years ago to form 750 billion galaxies which virtually all rotate along a central black hole. The cosmologic constant Λ *is based on the antigravitational effect of the expansion of galaxies in space-time and was discovered by Albert Einstein.*

Space-time=space/mass and from Einstein's Equation of Relativistic Gravity R ijk-1/2R g ijk=c^4/8 $\pi G(\wedge g\ ijk)$ *and from The Equation of Everything* $8\pi T\dfrac{ijk}{R}ij(\wedge gijk)$ *where R ij reflects the Ricci Tensor of inertial mass and $8\pi T\ i\ j\ k$ reflects the stress energy tensor*

Tensor. $\mathbb{R}ijk - \dfrac{1}{2R}gijk = 8\pi Tij\dfrac{k}{R}ij(\wedge gijk)$ *where R i j k and R g i j k are the Riemann forces. Weyl^' sConformal Tensor of gravity*

In cosmic inflation is as such C I j k=R I j k-1/2R g I j k=8

$\pi Tij\dfrac{k}{R}i\ j\left(\wedge gijk and\wedge = \dfrac{c^4}{8\pi G} = \dfrac{1}{R^2 reflecting}\right.$ *the reciprocal curvature of space – time with Dark*

Energy As $\dfrac{c^4}{8\pi G}$ *is reciprocal curvature from Dark Energy* $\dfrac{c^4}{8\pi G} = mc^2$ *for Dark Energy yielding* $\left.\dfrac{c^2}{8\pi G} or \dfrac{3x10^{8m}}{s}\right)^2$

$\overline{33\left(6.67x10^{-19}\right)}$

22.9x10^32 kg as the mass equivalent of Dark Energy where R=radius of space-time curvature.

The missing mass mentioned earlier in this chapter was 3.3x10^34 kg in a universe 10^54 kg which is 2.29x10^33 kg or .029x10^34 kg which is only 100 kilograms away from the mathematically predicted value. THEREFORE THE MASS EQUIVALENT OF DARK ENERGY IS BETWEEN 0.229 AND 3.3x10^34 kg IF THE MISSING MASS AND MASS EQUIVALENT ARE THE SAME LOGIC CONCLUDES WE'RE DEALING WITH THE SAME OR LINKED PHENOMENA. IF DARK ENERGY AND DARK MATTER ARE TWO DISPARATE SUBGROUPS OF LARGER GROUPOIDS THEY WOULD BE ABELIAN IF DIFFERENT IN FUNCTION BUT IF SIMILAR OR THE SAME THEY WOULD BE NON-ABELIAN. THE METRIC OF SPACETIME CURVATURE INTERACTED BY THE METRIC OF DARK ENERGY AND/ OR DARK MATTER WOULD BE ANTISYMMETRIC WHILE THE RICCI TENSOR WOULD BE SYMMETRIC. $\mathbb{R}4$ *OR 4-BRANE*=$\mathbb{R}ABCD$=-$\mathbb{R}DCBA$ while Rab=-Rab=R ba with the Ricci tensor. The subgroup of of the groupoid with Dark Energy and Dark Matter with the missing mass equivalent is based on $\mathbb{R}ijkl$=Ri j k- $Rijk - \dfrac{1}{2R}Rgab$ *all divided by R ab which is The Equation of Everything where* $\dfrac{1}{2R}hasR = \wedge gijk$ *which is antigravitational effect on space-time from the cosmologic constant which is* $\dfrac{1}{R^2}as\dfrac{R^2}{R} = R.$

$$\frac{c^4}{8\pi G} = mc^2 = \frac{c^2}{8\pi G} = \frac{1}{R^2 w}h\frac{ere\left(3x108\dfrac{m}{\sec\sec}\right)2}{33xx10^{-19}} = 2.9x10^{34}\,gmor.029x10^{34}\,kg$$

QUERY. COULD DARK MATTER BE THE 'BURNED OUT'RESIDUAL OF ANTIMATTER AFTER THE BIG BANG WHERE THE ANTIMATTER ANTIMATTER REPULSION TRIGGERING THE EXPANSION FROM THE QUANTUM BUBBLE INSTEAD OF LEAVING A PAUCITY OF ANTIMATTER

(WHEN THERE WAS AS MUCH ANTIMATTER AS MATTER IN THE QUANTUM BUBBLE LEFT DARK MATTER+DARK ENERGY FROM THE ANTIMATTER AND THE RECORDABLE(VISIBLE) UNIVERSE FROM THE MATTER BEING PUSHED OUTWARDLY BY THE DARK ENERGY. AS A CHEMICAL EQUATION ANTIMATTER+ANTIMATTER →*DARK MATTER+DARK ENERGY WHERE DARK ENERGY REPELS THE MATTER CLUMPS AWAY FROM EACH OTHER*

OTHER AND 'THE BURNED OUT' DARK MATTER ACTS AS COSMIC GLUE FOR MATTER CLUMPS WITH POSITIVE GRAVITY WHILE THE ENERGY RELEASED WAS THE ANTI-GRAVIATIONAL EFFECT FROM THE MUTUAL REPULSION OF ANTIMATTER FROM THE QUANTUM BUBBLE. THE BARYONIC PARTICLES MAY HAVE ANTI-BARYONIC PARTICLES AS WITH NEUTRINOS AND ANTI-NEUTRINOS. IF THIS IS TRUE THEN DARK ENERGY AND DARK MATTER HAVE THE SAME PHENOMENON IN COMMON MAKING THE FUNCTION OF THE SUBGOUP NON-ABELIAN AS IN A GAUGE SYMMETRY GROUP OF YANG MILLS IF ONE CAN ISOLATE SUBATOMIC PARTICLE SETS WITHIN THE SUBGROUP WITH THE SAME OR SIMILAR BEHAVIOR. THIS TOO CAN ACCOUNT FOR THE MASS GAP IN YANG MILLS THEORY $\Delta>0$. *Not to regress to Gauge Symmetry Groups* the Ω-*particle in the SU(3)subgroup may relate to Dark Matter with the behavior of burned* out antimatter with positive gravitational effects while the Dark Energy mass would give the same mass gap as they both relate to the same phenomenon. String Theory may have already been incorporated in Yang Mills Theory with the SO(32) gauge symmetry group for string theory and photons were already incorporated into the U(1) subgroup as electromagnetic radiation consists of photons.

CHAPTER SIXTEEN SEVENTEEN (CONTINUED)

Using the Equation of Everything with R ab=inertial mass of the quantum bubble R ab= ρ 0 *where volume is* 10-33^{cm} (7.5x10^{11} *strings)*=7.5x10^{-22cm}.

R ab=ρ 0(7.5x10^{-22} *cm) so* $\mathbb{R}abcd=\mathbb{R}abc-\frac{1}{2Rg}ab \div Rab$ for matter and $+\frac{1}{2}Rgab \div Rab$ *for antimatter.* $\mathbb{R}abcd=\mathbb{R}abc--\frac{1}{2}Rgab \div \rho0\left(7.5x10^{-22}cm\right)$ yielding ρ 0(7.5x10^{22} *cm)* $\otimes Rabc-\frac{1}{2}Rgab$

ρ0=vacuum energy density of matter→0 for vacuum state

R a b c-+R g a b= ∞ *where±R g ab=∞ and R abc→0 yielding infinite curvature of contracted space-time R abc yields a point of infinite curvature for Rabcd from±R g ab=∞ acting on R abc→0 for flat contracted space*

In terms of the mechanism of "The Big Bang" with subcomponents of swirl and anti-swirl when the quantum bubble forms a double torus and then hyperbola will matter and antimatter sequester at the poles or extremes or will a homogeneous mix of matter and antimatter rotate to the poles like a centrifuge? As the observed universe is isotropic and relatively homogeneous with homogeneous background microwave radiation(BMR)one would have to conclude that matter and antimatter will intertwined and annihilated each other will the exception of the residual matter from matter and the dark matter from the antimatter plus Dark Energy. The central locus of the blast with pushing out of residual matter, dark matter and residual energy from antimatter-matter annihilation is what this author coined the(0,0) point in first book "Megaphysics, A New Look at the Universe".

In terms of mechanism polar coordinates of a radius r are x=r c is θ *where cis=cos+i sinθ and y=r cosθ-i sinθ. With regard to quadric surfaces the hyperbola of a flat sheet is* $\frac{x^2}{a^2}+\frac{y^2}{b2}+\frac{z^2}{c^2}=1$ *and the hyperbola of two flat sheets is* $\frac{x^2}{a^2}-\frac{y^2}{b^2}-\frac{z^2}{c^2}=1$ *which is related to the two dimensional world sheet with open or closed strings. In the case of* matter and antimatter the vector form (Ar).r+2a.r+a 44=0 where the tensor A has components Ai^k =aik as in the second degree equation $a11x^2+a22y^2+a33z^2+2$ a12xy+2 a13xz++2a 23yz+ 23yz+2a24y+2a34z+a44=0. $\mathbb{R}2\rightarrow4$ *in the quantum bubble is suggested by the vector equation (Ar).r+2 a.r+a 44=0 with regard to x,y,and z coordinates for a ik=a ki and [i,k]=1,2,3,4*

The space time curvature invariants of R g are I,J,D,A with respect to the metric g. I=a11+a22+a33 $J = \begin{bmatrix} a11 & a12 \\ a21 & a22 \end{bmatrix} + \begin{bmatrix} a22 & a23 \\ a32 & a33 \end{bmatrix} + \begin{matrix} a33 & a31 \\ a13 & a11 \end{matrix} det$

$D = A44 \begin{matrix} a11 & a12 & a13 \\ a21 & a22 & a23 \\ a31 & a32 & a33 \end{matrix} det$ and $A = 4x4$ *det aik* $\otimes aki$ where i=1,2,3,4 and k =1,2,3,4

A'=da/dt=A11+A22=A33+A44 and A"=det a ik $\otimes a$ *ki with* A'''=a11+a22+a33+a44. Korn and Korn 3.51-Mathematical Handbook for Scientists and Engineers

There will be a Feynman Diagram showing the rotational vectors at the extremes of the quantum bubble in opposite directions CW and CCW or for closed bosonic strings left and right handed components representing the mass operator m2. It will show a flat one dimensional surface(Riemann surface) with cw and ccw rotations turning into a two dimensional surface. With Clockwise and counterclockwise rotations approaching the extremes of the poles in the form of a double torus they turn into a hyperbola with clockwise and counterclockwise rotations at the extremes which were initially the poles in an osculating circle where unit tangent vectors touched or osculated the surface of the circle. The mechanism of the 0,0 point with the symmetrical throat and position vectors will be discussed again later with Feynman diagrams.

CHAPTER SEVENTEEN

SPACE-TIME BOUNDARIES, THE BIG CRUNCH AND THE SPEED OF LIGHT

If greater than a specific mass of this universe was traveling near the space-time boundary of 3×10^8 meters/sec(c) space-time would start to rip or tear the way a pair of pants with a small rip increases with pressure. This in turn could create a "Big Crunch" WITH A SLOW PHASE WHERE THE RED SHIFTS FROM DOPPLER WOULD START TO SHIFT TOWARD THE ULTRAVIOLET MAKING THE ACCELERATED EXPANSION OF GALAXIES SLOW OR BEING DAMPENING followed by a fast phase where the collapse of space-time and the 750 billion galaxies would begin to coalesce at a faster rate. In these instances as an alternate to Heat Death" mentioned previously there would be time for intelligent observers to prepare. Preparation would have to be time or dimensional travel which is beyond our present technology. If on the other hand there was cosmic inflation with a "pop" like a balloon puncturing the "Big Crunch" could take 10^{-43} seconds or Planck Time and most intelligent observers would not even be aware it happened. Luckily, even with the Lorenzian Transformations a small enough percentage of total mass in this universe is traveling at or near the speed of light according to the de Broglie Equation $\lambda = h\frac{}{mv}$ *where v→c that that threshold is far from being met, which would make heat death a more likely scenario*

Where everything eventually decelerates, almost stops and temperatures from to the vacuum temperature of space of 2.74 degrees kelvin.

What is a boundary. A boundary is an interface between states or groups of states being acted upon by an operator. Dr. Steven Hawking stated that the "only boundary is that there are no boundaries". If he is correct singularities should not occur where the laws of physics break down. General and Special Relativity say no matter what the reference frame the laws of physics must remain intact making a Big Bang ex nihilo impossible as it violates Newton's Third Law(every action has an equal but opposite reaction)and The Laws of Conservation of Energy and Momentum. In the case of a Big Crunch the 2nd Law of Thermodynamics stating that S(entropy)will always got from a more to a less ordered state would possibly be reversed assuming the contents of a quantum bubble from a dimensional and galactic collapse is more ordered that the expanding universe it collapsed from. In a black hole also mentioned previously the entropy was postulated at approximately0.29 by Dr. Hawking while Albert Einstein postulated it as near infinity. The equation $S=2N(Q1Q5)^{\wedge}1/2$ where N=number of states of a black hole indicates a small number for black hole entropy which would mean a similar scenario in a Big Crunch. In addition Steven Hawking postulated that in a Big Crunch times arrow would be reversed which would justify a Big Crunch prior to The Big Bang where time went from positive backwards toward zero without reaching it as an asymptotic function and then after the Big Crunch would advance again forward. It time had actually reach zero instead of approaching zero it would have been a boundary. If matter whose inertial mass advances toward infinity would actually reach infinite inertia at "C" instead of approaching "c" that would also be a boundary. A if D-0-branes actually represented a zero dimensional state instead of a near infinite number of quantum dots compactified to zero dimensions that would be the boundary of space-less ness.

As one second of arc can be subdivided down toward an infinite number of cuts with the circumference of a circle being space-time the zero dimensional state is actually never reached per $s=r\,\theta$. The equation circumference$=2\,\pi R$ *of a circle which was compactified form of typeIIA string Theory and M theory relates to*

Black Hole entropy $S=2\pi(NQ1Q5)^{1/2}$ where radius$=(NQ1Q5)1^{1/2}$ and Q1 relates to the one-brane relating to the effect of the charge of the monopole or electron and Q5 relates to the 5-brane relating to reducing space-time. In a black hole space-time$=2piR$ has R reducing to near zero(0)at the central locus so space-time reduces to almost 0 at the central locus. S(entropy)reduces to almost 0 as $2pi(NQ1Q5)^{1/2}=2\pi(252$ different states)(1-brane)(5-brane)$^{1/2}$ approaches 0 as the 5- brane approaches 0 for space-time making the number for black hole entropy of 0.29 accurate. This will be discussed again later in Chapter 18 on black holes. The entropy of space-time is 0 if space-time contains no contents. If expanding and rotating does space-time have an entropy when it doesn't before the "Big Bang"? As space-time has contents after the Big Bang" and indeed before the" Big Bang" there must be a positive entropy in both cases. Does the entropy increase after "The Big Bang" for space -time? As the entropy of the contents increases after "The Big Bang" and the contents are a subset of the groupoid of space-time then the entropy of space-time must increase after "The Big Bang". The circumference of $2(\pi)R$ where C-space-time shrinks to zero as the radium or Reimann forces shrink to zero. It therefore follows that the entropy of space-time shrinks to zero as the radius shrinks to zero. With regard to commutative algebra, it can be shown that the vacuum state=everything. Assume $I \otimes A \frac{A}{J} \to \frac{A}{J}.(I+J)/J$ indicates A/I $A/I \otimes A \frac{A}{J} = \frac{A}{J}$.

$A/J \div \left\{ \frac{I+J}{J} \right\} = \frac{A}{I} + J.C^* =$ *envelope of certain quotients of tensor algebras over a C*correspondence. Let* $Rabc - \frac{1}{2R} gabbel + J$ *and R ab be* $J.Rab = \frac{1}{2R} gab$ *from Einstein^' sEquation of Relativistc Gravity as G ab=0 so it follows that* $I \otimes A \frac{A}{J} \to \frac{A}{J}$ *and* $I + \frac{J}{J} = \frac{A}{I} \otimes A \frac{A}{J} = \frac{A}{J}$. *Let* $I + j = Rabc - \frac{1}{2R} gab$ *and R ab=J then* $I + \frac{J}{J}$ $Rabc - \frac{1}{2} Rg \frac{ab}{Rab} = I \otimes A \frac{A}{J}.I + \frac{J}{J} = \frac{I \otimes A^{\frac{A}{J}}}{J} = \frac{A}{J}$.
If $\mathbb{R}abcd=0$ *then* $I + \frac{J}{J} = \mathbb{R}abcd = 0.Rabc = 0$. *R a b c=0 and when R ab=0 then R g ab=0 as space-time is flat in a vacuum then in the vacuum state* $0 = \frac{0}{0}$ *or everything. Based on the premise that nothing doesn^' t exist the vacuum state state is a TRUE*

TRUE BOUNDARY. The vacuum state is beyond the comprehension of mankind however as in this case can be calculated mathematically. Riemann surfaces apply to partial applications of Riemann where the forces $\mathbb{R}f$ *get applied to curved curved Riemann space such that R ab varies with each relationship*

Of R a b c d and R g ab as g ab varies with R ab(Ricci Tensor) in interaction with with Riemann space to form the space-time curvature variant R g ab. Electromagnetism apply as photons have a non-resting mass, light is bent by gravity and black holes do not allow light to escape plus the deBroglie equation which says that each and every mass has a corresponding wavelength and therefore frequency. Also the non-resting mass of a photon should increase at just below v=c by the Lorenzian Transformation m=m0/1-v^2/C^2)^1/2. However the observer must be traveling at be traveling at the same or nearly the same speed to detect it.. Therefore the Ricci Tensor for photons would approach zero (0) in most reference frames and the space-time curvature variant would approach ∞ *in contracted space-time causing space-time to contract to an infinite curvature point which is why space-time acts as TRUE BOUNDARY at velcoity=c or* $\frac{3x10^8 \, meters}{second}$ *and space-time approaches 0 as an asymptotic function at v=c.*

In the case where space-time as Riemann 4 space doesn't exibit much curvature such as the case of 5x10^-29 meters in a 10^54 kg universe $\mathbb{R}abcd=R\ abc$ *where R g ab=0 then Rabcd=R abcd are space-time is asymptotically flat such as Riemann 4 space=Mintkowski 4 space as in the cases of of everything except black holes and the quantum bubble* and acute tears of space-time as in the fast phase of a "Big Crunch".

Space-time curvature is maximal at or near the event horizon of a black hole or the quantum bubble or at or near "c". Most of space has very little mass besides the BMR the background microwave radation from the Big Bang which heats space to 2.74 degrees kelvin. This BMR is generally homogeneous according to WOMP and do to the lack of significant mass of the BMR does not curve space-time in the vast region of this universe significantly. The sum total of the 750 billion galaxies and numerous black holes in each galaxy still dwarf in comparison to the voluminous space that exists where mass is significantly lower. The sum curvature of all the black holes in Schwarzchild space-time and that of each galaxy will still dwarf in comparison to space-time (which is asymptotically flat) assuming no measurement error for the mass of the universe of 10^{54}kg. The vector product of $A \otimes B$ *is* $Rabc-0.5Rgab \otimes R \, ab^{-1}$ *scalar trace* is $R-1/2Rg \otimes R'^{\wedge}-1$ yields $\mathbb{R}$ *where the vector product* $A \otimes B = AB \cos\theta$ *where* $\theta = \pi$ *radians so* $\cos\pi = 1$. $R-1/2Rg \otimes R'^{\wedge}-1$ *as the* scalar trace has $\cos\theta(R \, a \, b \, c - \dfrac{1}{2R} \, ab \otimes R \, ab^{-1}$ *as the vector or tensor product where* $\theta = 1$ *as cosine* $\theta = 1$ *and* $\theta = \pi$ *radians*. With regard to space-time curvature from electormagneitic radiation start with the stress tensor of a perfect fluid (assuming space-time as a perfect fluid) $Tab = \rho\mu a\mu b + p(g \, ab + \mu a\mu b)$ *where the derivative operator* ∇^{aT} *ab=0. In the Klein Gordon Scalar Field with curved space-time we get what's called minimal substitution* $\eta ab \to g \, ab, \partial a \to \nabla^{a}$ *such that* $\nabla^{a}\nabla^{a}\emptyset - m^{2}\emptyset = 0$ *where* $\emptyset$ *is the scalar field m is the mass and* ∇ *is* *the derivative operator. T $ab=\nabla_a\phi\nabla_b\phi$-0.5 $gab(\nabla c\phi\nabla^{\wedge}c)+m^{2\phi^2}\rightarrow\nabla^a$ T ab=0. The electromagnetic stress tensor is based on thethe Maxwell Equations ∇^a F ab=$-4\pi j$ b where the dual vector field from Poisson$^{\wedge}$' sequation is F ab and j b relates to intensity*

While F ab relates to emf(electromotive force which would relate to voltage while jb would relate to amperage. The derivative operator of [AF bc] or $\nabla[aFbc]=0$. *The stress energy tensor of the field is*

$$\frac{1}{4\pi\left\{FacFb^c - \dfrac{1}{4g}abFdeF6de\right\}} \text{ or } 4\pi^{\wedge}-1.(FacFb^c - 0.25gabFdeF^{\wedge}de$$

With commutation of derivatives $\nabla^{a\nabla aA}b - Rb^{dAd} = -4\pi jb.$. *F ab is the elctromagnetic field or emf and the Lorentz force equation states $a^b = \mu^{a\nabla a\mu^b and}$*

q/m F^b $\mu c^{\wedge}c$ *from the acceleration $a^b=\mu^{a\nabla a\mu^b}$ and) of course ∇a is the derivative operator associated with g ab. These indicies* relate to F c^b by the metric g such that F c^b=g ^b d F dc. Basically electromagnetic fields as per the Maxwell Equations do have an effect on space-time curvature based on the Klein Gordon Scalar Field and the electromagnetic stress tensor T a b where mass is involved from the m^2 *ϕ^2 part of the equation.*

Stress energy relates to space-time curvature via 8 πT ab= $Rab - \dfrac{1}{2}Rgab$ *where R g ab relates to spacetime curvature variant of metric g ab describing the* effect of gravity and relates to electromagnetic fields via Maxwell's Equations and the Stress Energy Equations above.24

One columb is 6.241 x10^18 electrons and 6.241x10^18 protons=1 ampere/sec. One columb=one farad x one volt. Coluomb's Law

F=kQ1Q2/r^2 where k=8.988 x 10^9 kg/m^3/sec ^2 coulombs^2.Amps/ voltage=conductance in units called siemens/The aether(background quintessence or substance in which energy and matter interact) contains a conduction constant C d=kc/c $\frac{\mu0}{\varepsilon0}$ =2.11231939x10-4 *siemens. and k c=c C* $\frac{d\mu0}{\varepsilon0}$ *=coulomb's constant.* Cd=condictive constant c=conduction. Based on these equations and the mass of an electron=9.1095x10^-28gms=0.511 Mev/c^2 there is mass involved in current which curves space-time. Voltage or electromotive force involves the rate of change of an electron stream from the negative pole to the positive pole of a conductor. Photons are the fundamental unit of light or radiations. The energy of a photon E=hv *where v is the frequency of a photon. Photons are considered immutable or virtually indestructible*

-able and have a length lifetime. Photons add mass to a system eventhough considered by most physicists massless. The energy of a body=mc^2+0.5mv^2 involving the special theory of relativity incoprorates K.E.=mv^2/2+mc^2. Note that E=mc^2 is a low speed approximation but as a particle or packet approaches "c" E=mc^2 has to be adjusted. Therefore the energy of a photon^2=p^2c^2+m^2c^4 where p=momentum and m is the resting mass. Photons have momentum E=pc in the above equation if 0 resting mass is assumed at v→c *in our reference frame where we are moving at a considerable slower rathe than the photon* although special relativity states that no matter what the reference frame "c" the speed of light is a constant. However does this mean that photons would still be massless(resting mass)if the observer was traveling at nearly the same speed ?In the

Harvard Gazette in a 2007 article "Light and matter united" when a light pulse disappeared from one cloud and appeared in another cloud nearby so light and matter were interconnected or entangled such that the information in the photons was translated into the matter and then converted back to light. In a way this can be "The Law of Conservation of Mass" where mass and energy can be entangled or switched back and forth. It was done by Dr. Lene Hau who slowed photons to 38 miles/hour when atoms were packed at a superhigh density at super low temperatures the atoms act like a single super-atom in what appears to be similar to the state Boso-Einsteinian Condensate where the single super-atom shows laser like qualities. In this medium light or photons are successfully slowed to 38 mph. This experiment done at the Rowland Institute for Science in Cambridge Massachusetts indicates that photons or light may under certain circumstances have changes in mass which naturally would also curve space-time. So then what is "c"? The speed of light constant IN A VACUUM IS A FIXED VALUE OF 2.99x10^8 meters/sec and much of space is nearly a vacuum however a true vacuum state as mentioned previously is without space or spaceless which would be the state of anything at the boundary of c. Are photons truly massless? The mass has been postulated at, 1x10^18 ev/c^2 with a charge q,10^-35 electrons with a spin of 1 charge of →0(*as mentioned*) *parity of-1 and c parity of-1 giving the term* $1(J^{PC})=0,1(1—)$ *with no resting frame and again as mentioned before* $E^2 = p^{2c^2} + m^{2c^4}$. *Motion is+or-ħ as also mentioned previously in a helical motion such as DNA. Parity is the reflection of the coordinate system through the origin* ordinate system through the origins in the CPT theorem whose operate commutes with the Hamiltonian operator. Force laws must be unvariant or unchanged subjected to charge conjugation and the CPT theorem works in any direction of time so CPT=TPC,,=PCT. Lorentz Invariance must follow the CPT theorem in Quantum Field Theory including charge conjugation, space inversion and time reversal... Based on this the c parity and parity of -1 reflects a negative reflection of the coordinate system including the inversion or reflection with regard to space. At or near black holes do photons get pulled in by the mass of the black hole curving space-time down to a contracted point or is the miniscule mass of a group of photons attracted by the extreme gravity effect of the black hole? More physicists should be agnostic regarding whether or not photons have a mass as there are specific situations which imply miniscule mass although for all intents and purposes photons are still massless with regard to resting mass. Do photons curves space-time or do they simply travel the helical path along already curved space-time?

Another boundary is ABSOLUTE ZERO OR 0 degrees kelvin. It is postulated that at this boundary are motion with a system totally stops which would approximate "Heat Death" in local regions of our universe. Matter and energy act differently at or near absolute zero as mentioned in Dr. Hau's article. If Boso-einsteinian Condensate contains super atoms and photons trapped inside the condensate at superslow speeds could they achieve different states the way black holes achieve 252 different states of matter according to Steven Hawking? Can radiation be liquefied as queried earlier under extreme pressure and near absolute zero? If it can and if somehow radiation and matter can be interchangeable or entangled radiation may classify as a perfect gas(as mentioned previously)as some have called space-time a perfect fluid. Could this also lead to transporter devices which today is more science-fiction than fact.

Are membranes boundaries ?This point would be up to conjecture as there is not enough empirical evidence to either support or refute this question If a system is bound by a membrane the term "bound" implies a boundary. If a boundary can be breeched is it still a boundary? Boundaries adjacent to manifolds can theoretically be breeched through the 11th dimension according to super gravity and electromagnetic effects can cross membranes theoretically as the electron or monopole's effects manifest in the 1-brane while the monopole's causes are in higher branes. Lisa Randall postulated that gravity may leak from another universe or universes through the 11th dimension to this universe which is why gravity is a weak force in this universe or manifold. Discovery channel special on Dr. Randall. and string theory.

CHAPTER EIGHTEEN

BLACK HOLES AND RELATIONSHIP WITH DIFFERENT STATES OF MATTER AND BLACK HOLE ENTROPY

Matter and energy are entangled under some specific circumstances. Are strings flat matter which is 2 or possibly 1 dimensional or are they 0 dimensional as energy only with motions tension and vibrations as only frequencies. Is there a Law of Conservation of Matter as matter and energy are inter-changable and entangled under the extreme pressures and temperatures in a black hole? According the Schwarzchild Space-time there is a constriction of space-time with time dilating to almost infinity or in essence stopping at the Event Horizon of a Black Hole. Black holes emanate from collapsing matter in galaxies, neutron stars and possibly universes. As a consequence there should be a black hole at the 0,0 point which is the point of the "Big Bang" although in an isotropic universe scientists may not be able to locate it for a considerable period of time The initial rotational vectors in the expanding universe where rotation is progressively decreasing as space-time continues to push outward should be slightly measurable as the proximity to this 0,0 black hole is approached although iso-tropism generally precludes a center of gravity. Still it makes sense that the 0,0 point would be a center of rotation for the rotating and expanding universe(this universe in the multiverse)It is now postulated that all galaxies have a central black hole including the Milky Way and these galaxies rotate along the central axis of these black hole albeit at an extremely slow rate. If negative mass exists in a black hole it was mathematically determined in chapter of Schwarzchild Space-time that superimposition of region 3 on region 2 and regions 1 and 4 can cause the information to garner at the Event Horizon like a phonograph record or video tape. This relates to the idea of The Holographic Universe and translates everything into two dimensions. Regions 1 and 3 are left and regions 2 and 4 are cancelled. Region 3 has negative mass and superimposes on region Region 3 is reflected back on region2 with region 3 having a negative mass and region 2 having a positive mass and there canceling as per the math in chapter 5, leaving regions 1 and 4. This may be a solution to the Hawking Paradox which says that as a black hole evaporates information in the black hole is lost.

Regarding the 252 different states of matter within a black hole(Hawking) these would have to include all or most state under extreme pressure with significantly constricted space and extremely high density. If electromagnetic radiation including photons from light are absorbed into a black hole(making it invisible)would that radiation show matter or matter like characteristics such as the states of liquid radiation suspended in a condensate as in the experiment by Dr. Len Hau mentioned in the previous chapter. If radiation could occupy the different states of matter and if matter and radiation can be or are entangled then radiation can occupy a liquid and possibly a Boso -Einsteinian Condensate state at near 0 degrees kelvin. If this is true perhaps radiation can in the future be considered a "perfect gas" but again for that one would have to demonstrate mass in a photon although of course electrons and photons have mass as do anti-protons and positrons although the mass in the latter two may be considered "strange mass" which might be an oscillating hybrid between negative and positive mass although recorded as positive mass which may or may not have anti-gravitational effects rather than the effect of pure gravity.

In terms of The Equation of Everything" as space-time curves in manner reciprocal to ordinary space or curves inward when in the expanding universe space-time curves outward $Rn=R\ abc-\frac{1}{2}R\ g\ ab$ $\div R\ ab$ or $Rn=\prod n=1to\infty\frac{1}{2^{n}\pi}g\ abR\ abc-\frac{1}{2}R\ g\ ab\div\rho\ ab$ where the infinite product of $\frac{1}{2^{n}\pi}$ times the metric $g\ ab$ times Mintkowski Space-time divided by the energy density of matter for the metric g ab reveals for a large mass

In constricting space reaches a point where $-1/2R\ g\ a\ b=R\ a\ b$ where the space-time curvature variant of the mass $R\ a\ b=$ inertial mass of the black hole such that $\mathbb{R}n=R\ abc\otimes1$ where $R\ abc$ constricts with near infinte curvature from$R\ g\ ab$ and $R\ ab$ is a huge number. The expression $\prod\frac{1}{2^{n\pi}}\rightarrow\frac{1}{n\pi}g\ ab=\frac{1}{\infty}=0$ space-time with near infinite curvature in a black hole with constricted space. Here $\mathbb{R}n\rightarrow0$ as n approaches a large number. The expression $\rho\ ab$ relates to $R\ ab$ as the enrgy equivalent of the inertial mass $R\ a\ b$ and incorporates the

$1/c^2$ or $1/p^2c2+m^2{}^\wedge c^4$=energy and p=momentum incorporates into *ρ ab the energy density with regard to the metric g ab. Of course the 1/2 π relates to the spherical nature and circumference 2πρ at the event horizon of a black hole where*

space-time is constricted down toward 0 with infinite curvature and as the areas where the energy density of matter *ρ and inertial mass constrict more and more it goes from* $\dfrac{1}{2\pi}$ *to* $\dfrac{1}{4\pi}$ *to* $\dfrac{1}{8\pi}$ *times the metric g ab and the stress energy*

stress energy T ab →1.*as inertia←gravity. As ρ ab→large number the entire expression for* $\mathbb{R}n$→0 *but where does the energy go from the event horizon of a black hole. Answer quasars with*

Hawking Radiation being spumed out like a jet engine leaving the black hole cold as a C02 cartridge would be cold after the contents of the cartridge were suddenly forced out. Therefore *ρ ab would manifest the energy for the inertial mass R ab like a CO2 cartidge being discharged over a* protracted time period.

The expression S=2 *π(NQ1Q5)$^{1/2}$ describes black hole entropy in terms of it's 252 different disparate states. This would be similar* in some ways to 2πρ=*circumference (space-time)*→0 *so ρ*→0 *deep within a black hole* and the diminishing sphere of space-time to a point is described by $\mathbb{R}n = \dfrac{1}{2^{n\pi}}n \to \infty\ gab \otimes Rabc - 0.5Rg\ ab \otimes \rho$ *as* ab^-1 where space-time= 2πρ. Note that a circle reducing to a point is a cone or asymptotically a spiral. Therefore the operator Πn=1 *to∞ ½^nπ would be the spiral operator on space-time reducing it* from a sphere to a point in a cone shape The Spiral operator would be $Ð\dfrac{1}{2^n}\pi^{-1}$ *where the infinite product* Π *is from n to* ∞ eigen-states.

The Bekenstein Hawking Equation for black hole entropy $S=A/4Lp^2c^3A/4G\hbar$ *where* $\hbar$=*Planck's Constant* $6.63x10^{-34joule}$*-sec or meters*^{2kg}*/sec G=gravitational constant* $6.67x10^{-11}$ *newton meters/ sec*^{2} *Lp=Planck length=*10^{-33}*cm* A=cross sectional area or kA/ Lp^2 where k=Boltzman constant and Pl=$\hbar\dfrac{G}{c^3}=10^{-33cm}$ *again. Black hole entropy (S) based on this is* $\dfrac{1}{4}$ *or also Steven Hawking postulated*

0.29. Black hole entropy is directly proportional to the area of the event horizon by the Boltzman Constant k which was explained earlier with the Boltzman Equation for different states of matter. This is the maximum entropy obtained by what's called the Berkenstein Bound and relates to the Holographic Universe and principle of a two dimensional fingerprint of information in the black hole at the event horizon. Supersymmetry was applied to black holes using D-branes and string theory duality with regard to the SO(32)string theory and the compactification of closed string theory IIa to a circle with regard to M theory.

The zeroth law states the surface gravity of the event horizon of a stationary black hole doesn't vary. The first law states that the the change in energy dE=k/8πdA+ΩdJ+ΦdQ *where* κ=*surface gravity A=area of event horizon*Ω=*angular velocity J=angular momentum* Φ=*electrostatic potential of the charge Q* The second law is that the horizon area is a non- decreasing function with regard to time or dA/dt> or =0. Whenit was discovered that Hawking Radiation was emitted by black holes and the area and mass(therefore gravity) decreased over time this was became a" weak law". The third law of black holes is that k *or κ for surface gravity cannot be 0 The zeroth law states that surface gravity is similar* to thermal equilibrium in thermodynamic systems in that a temperature doesn't vary neither does surface gravity *κin a black hole.* Wikipedia; Bekenstein Entropy

In the equation S(BH)=$2\pi\sqrt{NQ1Q5}$ N is the number of states as mentioned before and Q1 relates to the one-brane which relates to effecting or carrying the electrical charge related to the monopole or the electron emanating from higher states of matter in terms of n-branes.Q5 relates to the charge relating to space-time as the electron has a cloud with a probability density that goes out to ∞ *without reaching it but bounded by the event horizon of a a black hole.*

Based on this S=$2\pi\sqrt{(252)}$ *charge of an electron gas (space-time)*

Charge of an electron is 1.6x10^-19 coulombs and the 5brane relates to near infinite curvature of a contracted area of space-time as told by $\frac{\Pi1}{2n\pi}$ *g ab R abc*$-\frac{1}{2}$*r g ab*$\otimes\rho$ *ab*$^{-1}$ *where n=number of eigenstates in n dimensional space. for the infinite product from n to∞ giving spacetime*

½^nπ *as* $\frac{1}{2^0\pi}$ *as space-time approaches the D-0 state or D-0-branes* *making* $\frac{1}{2\pi}=\frac{1}{2\left(\frac{22}{7}\right)}=\frac{7}{22(2)}=\frac{14}{22}$ *so* $44/7\sqrt{252(1.6x10^{-19})\left(\frac{14}{22}\right)}$ =6.14[(160.36)(1.6x10-19)]^1/2 which is 6.141(6x10-17=36.6x10^-17 as black hole entropy in the zero(0)dimensional state or the D-0-brane which is very close to 0 entropy while Steven Hawking postulated entropy of a black hole to be approximately 0.29. This is because while spacetime is constricted at the event horizon it still exists so there is no 0 dimensional state within a black hole which would make the 5-brane relate to a four or perhaps higher dimensional state with the 252 different states of matter including energy-matter conversion or entanglement due to the superhigh pressures and super cold temperatures. It is likely that Boso- Einsteinian Condensate would exist with a matrix that traps photons, electrons, positrons, neutrinos, anti-neutrinos, bosons and at near the zero dimensional state fermions. It is also possible that tachyons would be trapped in a black hole which would reverse time's arrow and have a negative mass or strange mass hybrid between negative and positive mass. It is these tachyons that would cause the −mass in the superimposed region 3 onto region 2 that would solve the Hawking Paradox. Whether radiation actually takes on a liquid state is agnostic but not impossible under those conditions.

If the spiral operator operates on the function R a b c-1/2Rgab $\otimes \rho ab$^-1 and is inclusive of the metric g ab over the infinite product of eigenstates over n dimensional space it is asymptotic to Schwarzchild Space-time if the expression is reflected at the event horizon to express the increasing cone or spiral from approximately zero(0)dimensional space-time to n dimensional space-time where n goes from the D-0 eigenstate to D-n-eigenstate. This can be accomplished if $\prod$ *from n=11 or n=10 decends to n=0 for the D-brane which gives an opening spiral or cone past the event horizon as there is a descending cone to*

Cone at the event horizon with space-time spiraling in from ordinary space. The factor relating to the mass equivalent on both sides of the event horizon emanates from the function on which the operator is operating or "The Equation of Everything" so Schwarzchild spacetime=$\prod n \frac{1}{2^{n\pi}} - \frac{\prod n1}{2^{n\pi} g} ab$ *where the limits of the infinite product* $\prod$ *are from n to→D 11?and fromD11→n where n starts at D=1 not D=0 as the spiral is a modified cone acted upon by the metric g ab as per*

$Rabc - \frac{1}{2R} g\ ab \otimes (\rho\ ab)^{\wedge} -1.$

CHAPTER NINETEEN

DARK ENERGY AND DARK MATTER

It was questionable about whether dark energy and dark matter were related until the missing mass calculation for the mass equivalent of dark energy and the mass of dark matter were found to be congruous. As so they share charcaterisitcs and can be considered in the same gauge symmetry group. This can be negociated with the mechanism of the "Big Bang" which is like lighting a match. From a chemical standpoint(as previously mentioned)the unlit phosphors would be the homogeneous antimatter-matter mix, the Big Bang would be the light being struck, and the charred residual would be the dark matter while the energy expended would be the Dark Energy coupled with the energy from the annihilation of matter and antimatter in the "Big Bang".

As mentioned previously the antiparticle-antiparticle repulsion o\ in the quantum bubble under extreme pressure and temperature force a huge anti-gravitation force to push matter with space-time outward after the rotational component of "The Big Bang" had almost completely slowed toward 0. The anti-gravitational force pushing galaxies apart from each other is Dark Energy and the residual antimatter that was burned out is dark matter. As the anti-gravitational moment of the interaction was carried off by the Dark Energy, Dark Matter has a gravitational effect instead of antigravity although it emanated from anti-matter.

As mentioned previously the particles of dark matter are on and about everything just as the BMR from the "Big Bang" is all pervasive, but the particles of dark matter are so small yet homogeneous that they might be a multiple of $Pl = \hbar \frac{G}{c^3 \, or} 10^{-33}$ *cm which would make the density of this superfine powder or dust have a very* slight measurable density in the massive volume of space although the gravitational effects from the mass are considerable although measured indirectly. This too was already mentioned.

It has been postulated that Dark Matter is composed of baryonic particles and neutrinos possibly with anti-neutrinos and anti-Hadrons as a residual from anti-matter and that Dark Matter acts as a type of "cosmic glue" which has been present since the "Big Bang"

There is a question as to how much of the cosmologic constant Λ *is related to Dark Energy, but the force to push the* 750 *billion galaxies apart from each other* seems to be far greater than what was empirically measured for Dark Energy and greater than Λ *(the cosmologic constant) which relates to* $8\pi G/c^4$.

With a stretch of the imagination and some creative math one can see that the 1st,3rd,5th and all odd dimensions have dark matter sequestered as it's gravitational effect can only be indirectly measured and mass in the even dimensions $2^{nd}, 4^{th}, 6^{th}, 8^{th}$ etc. can have it's properties directly measured. This can evolve from the space-time formula of $\frac{1}{2}e^{\wedge}$-i n cot θ *where θ=π radians which is the trajectory of the "Big Bang"*. Odd powers of i give results containing I while even powers of I give real numbers such as $i^{\wedge}2$=-1, $i^{\wedge}4$=-1 etc while $i^{\wedge}1$=I and $i^{\wedge}3$=-I and $i^{\wedge}5$=-I such that if this was the scenario due to the inert nature of dark matter its mass may be postulated mathematically mass of dark matter=mass of ordinary matter/i *or* $\sqrt{-1}$. This would be likely if dark matter which is anti-matter based showed anti-gravitational rather than gravitational effect as Fg=Gm1m2/r^2 where 6.67x10-11n-m/sec^2 or 10-11=G and m1=mass/i m2=mass2/i such that Fg=Gm1m2/i^2r^2=-1(G m1m2/r2)which would indicate mutual repulsion of antiparticles and anti-gravity. While it may be true that the odd dimensions may have dark matter and possibly most of the missing mass equivalent of Dark Energy forming the difference between what is measured and mathematically calculated as a prediction on we will still have to show strange or oscillating mass in dark matter(as burned out antimatter). This oscillating property would be between + and – mass and while experiments with the Hadron Collider in Cern, Switzerland are looking for anti-gravity between anti-particles and particle-anti-particle interaction before they annihilate, it has not clearly been demonstrated yet that anti-particles display pure anti-gravity although they do display some properties of anti-gravity. It also has not been demonstrated that anti-particles display anything except a positive mass. The question is this. How does experimentally show burned out antimatter in dark matter when it's effects can only be indirectly measured. Also is mass/i the same as strange or oscillating matter with a hybrid between +mass and −mass. If this property follows antimatter and antiparticles it may also follow dark matter with much huger antigravity with Dark Energy. Mathematically shadow odd dimensions cannot be ruled out but whether or not they contain the mass of dark matter or mass equivalent of dark energy is up to speculation.

CHAPTER NINTEEN CONTINUED
DARK MATTER AND DARK ENERGY

Using Ockham's(Occum's)Razor which scenario is most likely in our universe with it's laws for our immediate reference frame? When unaccounted for mass is indirectly measured by gravity measurements and there is no other clear objective measurable evidence of it's existence is dark matter in odd dimensions of space-time possibly showing a parallel universe which has an opposite rotational vector in space-time (The Anti-swirl to the swirl of the rotational vectors of "The Big Bang" where space-time=+1/2e-i n cot θ or -1/2 e^in cot θ *where the sum total of all space-time curvature is* 1 *or asymptotic flatness or the dark matter and dark energy are biproducts* of "The Big Bang" where Dark Matter is the residual of the anti-matter and Dark Energy was the anti-gravity. Also to be added that the micro-particles or micro-dust of dark matter is so small and homogeneous that it's mass is hidden by the enormous(in comparison)volume of space. Obviously there is very little evidence of parallel universes but it can't be ruled out. However with our level of technology it should be provable that micro-particles of dark matter do exist especially with the Hadron Collider. How does one prove the strong anti-gravitational effects of Dark Energy when only weak readings are being recorded. Obviously something must be getting missed in the measurements although reproducibility implies precision, there may be some soft?f actors that are being underweiged or over weighed. The comes the third most sobering possibility which is probably unlikely, that dark matter readings from gravitational perturbations and the missing mass is due to measurement or reading errors due to some soft factors(Compton effect?)and the rate of the recession of galaxies from the Red Shifts might be off because of error in the speed of light due to incorrect readings of photonic mass. This author does not want to think that the third scenario is true though it can't be totally ruled out even twenty years or more since the hypothesis of dark matter was first postulated. Ruling out scenario #3 it seems most likely that dark matter is a byproduct of" The Big Bang" just as the BMR is and that there is a micro-dust adherent to virtually all mass as a burned out residue from antimatter while Dark Energy is the anti-gravitation force from the antimatter "burn out". Still, it is possible that odd dimensions in space-time reflect a parallel manifold just as CW manifold would be complemented by a CCW manifold and that dark energy and dark matter have most of the cause of the measured effects in those odd dimensions. Then there is scenario 4 which is that scenario's 1,2 and 3 are all wrong partially wrong or inaccurate. The scenario of the fine microdust also explains the mass gap $\Delta t > 0$ *in Yang Mills Theory with another non-Abelian gauge symmetry group with Dark energy and Dark Matter. This scenario would close the gap of the* Ω *the mystery of the* Ω-*particle in the gauge symmetry group SU(3)involving the Strong Force.*

Does the change of the existence of Dark Matter change with the presence or absence of an intelligent observer. In other words can "Spooky Action at a Distance" or entanglement be applied in this case as with the disappearing and reappearing of an electron(which has a gas or region extending out to other electron gasses or regions at vast differences with their exchanging information). This particular author is agnostic(questioning the validity) about the idea that Dark Matter appears and disappears whether or not an intelligent observer exists taking measurements. Although Schrodinger's Cat implies that this scenario could occur where an observer could have dark matter shift into the odd dimensions from the even dimensions and revert to the even dimensions when no intelligent observer is present. There is mathematical data to support this as there is mathematical data on space-time to indicate the slippage of Dark Matter from visible measurable dimensions to invisible indirectly measurable dimensions there is a paucity of objective evidence to support this hypothesis and it still isn't "the simplest explanation of the phenomenon" as per Occum's Razor. Still are all things equal? If not Occum's razor would not apply. Still "The Inter-connectiveness Theorem of Bohm" has an experiment shown on a Discovery Channel broadcast in the 1990's in which ink in a glass of water or solution swirled one way would made the ink disappear completely and swirled the other direction would made the ink reappear and start to dissolves. It was postulated that the ink was sliding to and from our measurable dimensions which can be observed to dimensions where measurement would have to be indirect and cannot be observed. Does this experiment support the idea that Dark Matter is slipping into odd dimensions which can only be indirectly observed and back to observable dimensions when there is no observer. If so there should be a counteraction to the action which made Dark Matter stay in the unobservable dimensions just as swirling the ink solution in the opposite direction makes the disappearing ink re-appear. If so what would be the counter-action ?Does the above make this scenario more likely ?The Inter-connectiveness Theorem states everything is connected or interconnected with everything else which would help explain "Spooky Action at a Distance". How valid this this assertion? The way this author sees it there isn't enough empirical data to either support or refute it and if there are alternative explanations to the swirling ink experiment that might refute it partially. However with Superstrings there may be a type of inter-connectiveness. within supersymmetry groups such as the SO(32)gauge symmetry group and overlapping gauge symmetry groups.

Dark Energy relates to a scalar trace of a field acting on a 5-brane with regard to a 4-brane as Φ where $\dfrac{d2\Phi}{dt^2}+3\dfrac{Hd\Phi}{dt}+V(\Phi)=0$ where $H^2=\dfrac{8\pi}{3m^2}\;\rho\left(1+\dfrac{\rho}{2\sigma}\right)+\dfrac{\Lambda 4}{3}+\mathcal{E}\dfrac{}{a4}$ *where the tension on the 4-brane from the 5-brane is σ and ρ is the energy density of matter whereE is an integration constant where* space-time curvature is acting on the 4-brane, The stress energy tensor relates to the Ricci Tensor via Einstein's Field Equation for Relativistic Gravity where $8\;\dfrac{\pi G}{c^4}$ $Tab=R\;ab-\dfrac{1}{2}R\;gab$ where the stress energy on the brane relates to the Tension or σ on the brane.

With regard to anti-gravity $8\pi\Lambda$ *can be substituted for-* $8\pi G$ *so* $-8\pi G$ *becomes* $8\pi\Lambda$ *and* $\dfrac{8\pi\Lambda}{c^4}$ *Tab=R ab-1/2(-R g ab) to show reciprocal curvature for anti-gravity* $=Rab-\left(-\dfrac{1}{2}\right)R\;g\;ab=R\;ab+\dfrac{1}{2R}g\;ab$ *so* $-8\dfrac{\pi G}{c^4}T\;ab=R\;ab+\dfrac{1}{2}R\;g\;ab$ $=8\dfrac{\pi\Lambda}{c^4}T\;ab=G\;ab=0.$

CHAPTER NINTEEN-TWENTY

The stress energy tensor of gravity is equal but opposite to the stress energy tensor of anti-gravity again substituting -8 πG for $8\pi G$ in the *Field Equation for Relativistic Gravity and applied to σ which was the tension on the* 4-brane. Negative tension would be a relaxation on the brane such that $H^2 = 8 \frac{\pi}{3m^2} \rho\left(1 + \frac{\rho}{-2\sigma}\right) + \frac{\Lambda 4}{3} + \varepsilon \frac{}{a4}$ *or* $\frac{8\pi}{3m^2}$ $\rho\left(1 - \frac{\rho}{2\sigma}\right) + \frac{\Lambda 4}{3} + \varepsilon \frac{}{a^4}$. *As* $\rho\left(1 + \frac{\rho}{-2\sigma}\right)$ *and* $\rho\left(1 + \frac{\rho}{2\sigma}\right)$ *or* $\rho\left(1 - \frac{\rho}{2\sigma}\right) + \rho\left(1 + \frac{\rho}{2\sigma}\right)$ *we get* $\rho\left(1 - \frac{\rho}{2\sigma} + 1 + \frac{\rho}{2\sigma}\right)$ *cancelling so* $\rho(0) = 0$ *and* $H^2 = \frac{\Lambda}{3} + \varepsilon \frac{}{a^4}$ *where* $\Lambda 4$ *is the cosmologic constant of anti-gravity in in* 4 *space or the effects of the graviton field curving space-time with steep potentials where the where the gradient* $\phi(r) = \nabla\phi = \lim\delta \to 0 \int_{s1} dA / \int_{v1} dV$ *and the gradient* $\phi(r) = \nabla\phi$ *for a scalar function* $\phi(r) = \phi(x,y,z,)$ *relates to* $\rho/2\sigma$ *as the scalar field rolls down the gradient of* $\phi(V)$ *where Kinetic energy* +voltage Vpotential= ρ *which relates to the energy density of matter. As* $\phi(V)$⇑*thegradient*⇑ *as does the anti-gravitational effect on the stress energy tensor making* $-\frac{8\pi G}{c^4}$ *T ab a larger negative number than* $\frac{8\pi G}{c^4}$ *T ab for gravity. As* $T ab = -T ba$ *with regard to membrane tension T ab<0 during the relaxation phase of the tension from the roll down phase* $\frac{\Lambda 4}{3} = \varepsilon \frac{}{a^4}$ *where* ε *transmits spacetime curvature onto the* 4 *brane. So* $\varepsilon \frac{}{a^4}$ *where a relates to* $\frac{a'}{a}$ *in the Hubble Expansion*

Hubble expansion coefficient as these equations were discussed in chapter 16 on Physical Cosmology. Consider T ^{a} b the stress energy tensor (contra-variant)for antigravity then T^ab>0 when antigravity applies.

In conclusion proving the existence of dark matter and dark energy requires showing that the anti-gravitational metric is directly related to the stress energy tensor reducing tension of the 4-brane from higher branes with regard to Einstein's Equation of Relativistic Gravity with reference to the cosmologic constant and H=a'/a. This would relate to the contravariant tensor if T ab=-T ba=T^ab where 8 $\frac{\pi G}{c^4}$ *is approximately* $8\left(\frac{22}{7}\right)\left(6.67x10^{-11}\right)\left(3x10-8\right)^4 = \rightarrow 0$ *so in the case of anti-gravity* $R\ ab - \frac{1}{2}R\ g\ ab = 0$ *and for anti-gravity the expression is manifested as reciprocal curvature of* $+ R\ as + + \frac{1}{2R}g\ ab$ *so R ab=-* 1/2R g *ab*$=-8\frac{\pi G}{c^4}T^{ab} \rightarrow 0 = Gab.$ *This is in conjunction with the low energy readings of Dark Energy which means*

The anti-matter anti- matter repulsion of "The Big Bang" is a very weak force as is gravity yet the repulsive force is related to the distance squared where the distance is very small so the anti-gravitational force goes up so a small energy level of Dark Energy manifests with extreme anti-gravity at Planck Time because the distance squared $\rightarrow 0$ *yet when measured Dark Energy is a very weak force that acts like s stronger force due to the compression of the anti* the compression of the quantum bubble pre-Planck Time. This also relates to Λ *as a very small number related to* $8\pi\Lambda/c^4$. The energy of 6.75x10^34 erg in the first second=annihilation energy of matter and anti-matter plus anti-gravity of antiparticle anti particle repulsion which is F=Gm1m2/-1(r^2) where r $\rightarrow 0$ *but may actually be Planck length*(10^{-33cm} *which would bring the force to Gm1m2* (10^{66}) *erg*

where 6.67x10-11(10^66)mass of the anti-matter=6.67x10^55erg(m1) (m2) if the inverse square law accurately holds in the pre-Big Bang quantum bubble. The energy=FS cos π *where* $cos\pi$= -1 *and antigravity is* $-\dfrac{1(Gm1m2)}{r^{2or}} - 6.67x10^{55}$ *newtons so* $\dfrac{6.75x10^{34}erg}{-6.67x10^{55}\,newtons} = -s$ *as s* $cos\pi = -1(s)$ *or the volume of space in which antigravity is propelled is the reciprocal of the masses* of two closed strings (1.03x10^-21 cm)the masses of two closed strings is dictated by the m^2 operator $\rightarrow 0$ *or masslessness so* 6.75x10^{34erg}/-6.67x10^{55} *newtons*$\rightarrow$(0)(0) yields a huge number for the volume of space in which the energy involved with the anti-matter anti -matter repulsion must act and explains why Dark Energy is still spreading out and pushing galaxies away from each other over a near infinite distance.

CHAPTER TWENTY

QUANTUM DOTS, ASYMPTOTIC FLATNESS AND THE 0,0 POINT

When is an open, flat expanding universe have curved and not flat space-time? Space-time flatness is approached but never reached doe to the decreasing rotational vector at the onset of "The Big Bang" corresponding with the expansion. Eventhough the rotational component slowed down geometrically at Planck Time and the expansive or inflationary phase became the dominant phase. If one were to take an infinitesimal slice of space-time expanding with an uncoiling rotational vector it would come out spiral although the bulk of the curving occurred right before the "Big Bang" in the quantum bubble. The total curvature is 5x10-29 meters for a 10^{54} kg universe which is such miniscule curvature it appears flat rather than curved in non-mathematical observations although it was proven to be curved(as mentioned previously)in 1921 by the solar eclipse experiment done on behalf on Einstein. At the event horizons of the possibly trillions of black holes in this universe space=time spirals down from a flat spherical region down to an infinite curvature point as per the spiral operator. As this occurrence and the mirror reversal of the spiral(asymptotic to a biconar region) can be multiplied by millions of black holes in the Milky Way alone and the central black holes at the center of each galaxy(750 billion galaxies) contribute a goodly percentage to the curvature of space-time all spiraling down to the event horizon. Still as space-time is curved more adjacent to a red giant than the earth and it's measurable so one would expect more curvature of space-time in the universe. Still to the intelligent observer astronomer using redshifts to determine distances for galaxies and assuming photons travel in a straight line(except in extreme proximity to black holes and stars)space appears to be very flat like the Friedmann type II open flat expanding universe giving asymptotic flatness(approaching flat space-time without ever meeting it) rather than Einstein's closed curved universe. This including a decreasing rotational moment from the "Big Bang" which earlier this author referred to as a swirl but the swirl component of the 360 degree orb blast leeched out and radiated out with the expansion causing either curved inflation(Alan Guth Ph D)with conformal space-time where the rotation was extremely dampened by the expansion as per the gradient and rotational curl equations. If trillions of black holes with millions in the Milky Way alone have the spiral operator reduce space-time to a point from an ever descending sphere which radiates out from expanding space-time and has weak perturbations from the spiral effect of other black holes making space-time curvature $-1/2e$-in $\cot\theta$ *where* $\theta=\pi$ *radians and n is the number of dimensions. Despite this as mentioned previously the volume* of space with only the BMR at 2.74 degrees kelvin far exceeds the space adjacent to heavy masses or black holes but space would have to be a googolplex or greater than 10^100 to have the curvature as such a small figure. In a "Big Crunch" the amount of space-time curvature would increase first slowly then dramatically as a state mimicking the pre-Big Bang quantum bubble is approached. As was mentioned there must be a mass threshold traveling at v →c *to weaken space-time enough to cause a Big Crunch.*

Quantum Dots are the smallest unites of space-time below Planck Length which is the approximately size of the orbifold or Calabi Yau manifolds(surfaces) in string theory and may comprise Hilbert, de Sitter, anti-de Sitter, and Fock space in Quantum Mechanics. As space-time can be subdivided down from one second of arc towards infinity space-time can be subdivided down to some lower boundary which may or may not be Planck Length(10^-33 cm)After that point space-time might break down into a lattice formation. This lattice formation would compose higher unites of space-time which are subject to the metrics of energy including mass that act upon it. Below this threshold energy may not be able to affect quantum dots individually but only as units. The n-branes of M Theory incorporate space-time may be comprised of these quantum dots. As the number of non-compactified dimensions have been mathematically determined as 26 and since non-compactified space goes down to Planck Length or smaller it has been determined that 26 is the number of dimensions although there are an infinite number of intersections of non-parallel planes emanating from the subdivision of one degree of a circle being osculated by metrics of the Riemann Forces. Based on the intersection of two parallel or non-parallel planes is necessary for a dimension there may be a huge number of dimensions. This would mean that the spacing between quantum dots would not be exactly the same as in the vacuum state and that some metrics are acting upon them to change the interspatial regions between quantum dots. These quantum dots must be considered a matrix upon which the quintessence or aether is acted upon by the metrics of Riemann. On a string level space-time curvature at 5×10^{-29} meters would be considerable even in a 10^{54}kg universe. If these quantum dots are virtually massless it could be seen how space-time would travel at or above "c" as initially propelled from "The Big Bang" or Inflation Theories as a figure previously determined as *πc with only tachyons which have a negative mass sqaured also traveling faster than* c.

These quantum dots act like pixels on a screen but blur out to a continuum of space-time as they are in motion(rotation and expansion). According to Zeno's Paradox the subdivisions of two objects becoming closer and closer to each other never touch because there will always be space between the objects and that space can be subdivided down towards infinity making it another asymptotic relation. Basically, if space-time can be subdivided down to quantum dots which are not continuous with other dots what is the space between the quantum dots ?It is possible that the space between quantum dots could be another confluence of quantum dots and that could continue down toward infinity or there could be aether between the quantum dots if aether can ever get a clear definition. Also spacelessness could have it's natural niche between these quantum dots where the lattice would be intact and it would not cause a singularity. An implosion has a lower limit even for spacelessness and quantum dots could be this lower limit in which space-time exists. It is difficult to comprehend which of these concepts may be true as space-less ness may only be possible in this set which may be the primary set of everything with {∞·,~∞.) *where. is the continuum of quantum dots over∞. This makes "The Axiomatic of Incompleteness by Kurt Godel* ness" go into question as a superset with everything and nothing seems to be complete. However if nothing is impossible(beyond the comprehension of man-kind)how can it be put into a set or equation? Nothing is mathematically possible and can be incorporated into sets as the null set{} but there may be a ground state where nothing is an integral part of the ground state.

The 0,0 point was coined by this author in previous book "Megaphysics, A New Look at the Universe"(2003)as the point in the quantum bubble from which the Big Bang" occurred. As previously stated matter gravitated to one pole of the quantum bubble and anti-matter to the opposite pole of the quantum with the 0,0 where 10^{54}(point at the center. The quantum bubble evolved to a double torus and then a hyperbola with the extreme rotational moments in opposite directions forcing the contents to the poles with extreme centripedal acceleration and force pulling the matter and anti-matter to the poles forcing a torus to a double torus to finally a hyperbola with the center of the hyperbola being the 0,0 point. The 0,0 point had a super-symmetrical throat which forced the anti-matter and matter from the opposite poles to mix in a homogeneous mixture then gravitated the matter-antimatter mix to both poles before the center of the hyperbola fractured from the energy of the anti-matter-matter annihilation plus the extreme anti-gravitational force of dark energy from the anti-particle and anti-particle repulsion all centering at this 0,0 point leading to an isotropic homogeneous universe with a paucity of anti-matter but a plentiful amount of dark energy which was the residue of the anti-matter and the Dark Energy which was propelled outward with anti-gravity due to the close proximity of all anti-particles in the quantum bubble according to the inverse square law. After descending into the throat of the 0,0 point the mixture moved to the opposite end of the quantum bubble or double torus which lead the vectors to move from the bottom of the double torus to the poles where the structure became hyperbolic. Whether or not the Robinson Congruence can be applied to this symmetrical throat at the 0,0 point is up to conjecture but the Robinson Congruence is like a curved torsed superhighway for subatomic structures in this case from the symmetrical throat to the poles of the hyperbola.

In an isotropic homogeneous universe how can there be a center of gravity at the point of "The Big Bang" ? Perhaps this universe isn't totally isotropic and homogeneous. If there is a super black hole out there where the "Big Bang" occurred and the 0,0 point emanated from perturbations of rotational vectors should be present with the bi-conar space-time (Schwarzchild) forcing more a of a spiral configuration to the cone according to the spiral operator. Proof of this rotation about that particular black hole will be empiric proof of the mechanism of "The Big Bang" as more of a huge swirl and anti-swirl preceding the 360 degree orb blast at Planck Time. It may take many decades to get this answer even with the Hubble Telescope and gravity cameras. Please note if the Big Bang were an explosion there should be a center of gravity for this universe if it is anisotropic and not totally homogeneous. If however Alan Guth's Inflation Theory is correct then a totally isotropic homogeneous universe would occur with no center of gravity as the moment of inflation would be evenly distributed throughout the quantum bubble with the anti-gravitational moment and matter anti-matter annihilation energy would be pushing spacetime outward at each part of the quantum bubble. Also if the 13.7 billion year old black hole evaporated, dried up and no longer exists the only footprint would be the rotational vector of space-time around the region of that black hole which would still be determined by gravity camera but not the Hubble. So proving or disproving the 0,0 point at the center of a hyperbolic mass with matter and antimatter at the poles with a center of gravity for this universe would be difficult to demonstrate empirically although nonhomogeneous measurements from the WOMP of the BMR might help but some gravity perturbation data is needed to determine if there is or is not a center of gravity. If inflation is totally correct(which is compatible with antiparticle antiparticle repulsion resulting in Dark Energy with it's huge repulsive force from an r^2 which is very small) and curved space-time inflated at $v>c$ one will have to agree with iso-tropism but if Inflation is the law and if $v \rightarrow c$ *is a true boundary where space-constricts* and time dilates then there may still be a threshold for matter and perhaps energy traveling at just under the boundary before space-time will either "pop" or tear resulting in a "Big Crunch". This was mentioned previously as was "Heat Death" but the question is this…" Will inflation make a "Big Pop" more likely as a "Big Crunch" that just a gradual ripping of space-time when the threshold mass is approaching "c". If so it is unlikely anyone will know it happened. A gruesome thought. Still, Inflation may be right and this threshold may never be met. It is very unlikely that this threshold mass will never be met because it will mean that the inertial mass of the ENTIRE UNIVERSE OF 10^54 kg. would have to be traveling at just under "c" when in reality $\lambda = \hbar \dfrac{}{mv}$ *or the DeBroglie Equation* indicates that only 6.63x10^-34/10^54(velocity) or 6.63x10-42/ λ of 10^{54kg} *c of mass* of (10^54)/1-v^2/c^2)1/2 must grow to almost infinite mass at v=c where λ *becomes infinitely long when relativity states that all matter will become infinitely short* at v=c and where infinite mass will shrink to 0 mass according to the Lorenzian Transformation. It seems to this author that infinite mass shrinking to masslessness and infinite length shrinking to 0 length for ordinary matter is almost impossible which is why it's a singularity. "Heat Death" from Physics and Immortality by Frank Tipler, Ph. D. is much more likely for this precious universe of ours.

CHAPTER TWENTY-ONE

SPOOKY ACTION AT A DISTANCE

The phenomenon of "Spooky Action at a Distance" relates to the exchange of information such as electron spin over great distances which may change when an observer is present. This phenomenon with regard to electrons may be caused by an electron cloud which doesn't dissipate over great distances meaning the fingerprint or signature of the electron's information could exist kilometers away or farther distance which might not seem possible with normal physics. A similar phenomenon can occur with photon pairing although photonic information with the wave particle duality can travel at just under "c" if photons have a non-resting mass and "c" if photons are massless. The pairing of information between photons could be because the wave property of photons could be analogous to the electron cloud of an electron and since photons are virtually ageless and immutable, information is also immutable must be preserved and can be exchanged between paired photons over huge distances via the wave property of photon via De Broglie's Equation $\lambda = \hbar \dfrac{}{mc}$ *where masslessness m=0 makes an almost infinite wavelength and super high frequency with very high energy*

From the Planck's Equation E(photon)= $\hbar\upsilon$ *where* υ=*frequency.*
Based on these equations very low frequency waves or photons
such as radio waves have more of a mass than high frequency waves
such as x or gamma rays. Massive objects also have a corresponding
wavelength regardless of the velocity of the massive object ($\lambda = \hbar \dfrac{}{mv}$
and the wavelength gets shorter as the velocity $\to$c.

The observed positions of electrons was not where they were supposed
to be due to electron cloud probability distributions of |.e E|> πm^2/
ln2 where e=electric charge(Q) and E- electric field and mass is
electron mass and probability of a generator penetrating a sphere
of diameter of z such that |z| leads to an asymptotic divergence of
probability with a continuity limit of varying field strength and the
definition of probability being violated by the tail of the distribution.
With regard to carrier densities, the density distribution of matter is
gc(E)=8 $\pi\sqrt{\dfrac{2}{\hbar^3}\dfrac{me^3}{2}}$ *where* $\hbar$=*Planck's constant and* $\sqrt{E-Ec}$ *for E>Ec*
has Ec as the conductioon band between the density between the
density of energy states and is where r=0 on the sphere upon which
the generating function acts where 2r=z. In this instance "g" is the
scalar metric not G. Discrete levels of Normalization for electrons
in the volume of an infinite diameter sphere can be summated into a
convergent system where the field strength and asymptotic behavior
of the probability distribution acts on all particle masses with carrier
densities as the infinite diameter of the sphere is forced to converge at
the circumference or $\pi|z|^2$ *where z=diameter and* $|z|^2$ *is the probability*
distribution. Σz=0 to z $\to \infty \left(\dfrac{4}{3\pi r^3}\right)\hbar\left(E-Ec\right)$where the volume of the

sphere is 4/3 πr^3 and z is the diameter with the infinite sum of the system.

Based on the wave function with Schrodinger's Equation the probability distribution of the electron field density with regard to the charge of an electron will diverge to ∞ as $z \to \infty$ until the volume is such that z converges with the circumference which must be at time=∞ according to relativity as the circumference is expanding faster than the diameter (z).

As the electron cloud distributions of many or all electrons in a field which can stretch out vast distances information on spin can be exchanged between electrons in the electron gasses that compose the field.

The Schrodinger Equation

$$\psi(E) = gc(E) = \frac{8\pi \left(\dfrac{2me^3}{\dfrac{2}{\hbar^3}}\right)^1}{2} \sqrt{E - Ec \int_{Ec}^{\infty} gc(E)F(E)dE}$$

Where n= carrier density f (E)=Fermi Dirac Proability Function
and n0= $\int_{Ec}^{\infty} \dfrac{8\pi \left(\dfrac{2}{\hbar^3}\right)^1}{2}$ or $8\pi\sqrt{2/\hbar^{\wedge}3}$ me^3/2 $\sqrt{Ev - E' + ef}$ E/kT dE where
k=Boltzmann constant exists for each specific state of an electron gas. The action(S) of the metric g for an electron with a carrier density of E v acting on the electron gas has a 100% probability with localization of the electron gas at infinite diameter z in the expanding sphere of space-time. The action S=c^8/12 $\dfrac{\pi GR1R2(-EvE)^3}{2r}$ where R1=p0 and R2=n0 and P0 is the probability of the carrier density

of E v as a state of matter=ρ0. With R1 as carrier density of state of matter acting on n0 of the electron the action

of the carrier density n has a probability of 100 % for the electron gas acting on *ρ the energy density of all states of matter as determined by the Boltzman Equation where*

K=Boltzman constant and T=temperature in degrees kelvin. Therefore there is a 100% probability that the carrier density of an electron gas will interact with the density of matter in a diameter(z) traveling an infinite volume and not converging with the circumference until the diameter and the circumference meet which is why electron gas paths must intersect and exchange information such as spin.

With regard to photons which are purported to be massless and traveling at "c" while electrons travel at approximately v=0.65 c photon pairing is much simpler to explain in terms of contracted space and dilated time. Information of a photon(as dictated by the energy level= $h\nu$) *travels vast distances* at "c" and the information may actually also change with or without an observer assuming that the temperature is 2.74 degrees kelvin or higher. To travel faster than this the photon would have to be a wave or electromagnetic radiation rather than a particle bending or warping contracted space to transmit the information. Photons and waves are supposed to travel in a double helix path burrowing spirals into space rather than in a strictly straight line(if photons were actually massless)Despite this the "Spooky Action(S)at a Distance" is still quite spooky and resembles the technique for a transporter device.

CONCLUDING REMARKS

The purpose of this book is to introduce to the scientific community mathematical adaptations of concepts regarding the five duel string theories and M Theory in terms of "The Equation of Everything" and applying this equation to Black Holes with the Spiral Operator to demonstrate it's congruence to Schwarzchild Spacetime. In addition this author has attempted to solve the "Hawking Paradox" using superimposed regions of Schwarzchild Space-time with negative mass from tachyons(replacing + mass bosons) to net out the information over dilated time as 0 so information isn't lost in a Black Hole. Also, the existence of Dark Matter has perplexed physicists for 25 years and in this work this author has attempted a new theory based on math that Dark Matter exists on everything as a fine microscopic powder with significant mass but in the volume of the universe has no measurable density making it invisible. In addition there was a purported mechanism that when anti-matter and matter either blew out with "The Big Bang" at the 0,0 point and inflated out with Inflation. Again consider light a match with the light being the Big Bang, the residue on the top of the match is Dark Matter, the force of anti-gravitational(push)is Dark Energy and the phosphors of the unlit match is the anti-matter matter mix. The residual matter after the anti-antimatter repulsion and anti-matter matter annihilation is the dark matter and ordinary matter after "The Big Bang".

The fact that string theory can be compactified(curled up)into a circle as IIa string theory and possibly M Theory lead one to realize that the equation of arc length s=r θ *can relate to spacetime and the Riemann Forces such that space-time accelerates and the Riemann forces curve spacetime as increments of θ from* 2n π *radians where the circumference of the circle(compactified IIa closed string theory and M* theory outruns the diameter of the circle which represent the Riemann metric in both directions with R*ab*=R^*ab* such that the radius is the + or – Riemann metric.

In this case as the circumference of space-time is 2πR *and since* 2R=d *then the rate at which space-time is" outrunning"* the Riemann forces is πc *as space-time and perhaps tachyons are the only quantities that can exceed the speed of light boundary.*

Concluding remarks continued.

The speed of gravity was also purported to exceed "c" in approximately 2005 A.D however as gravity is the curvature of space-time, it makes sense that the rate of curvature of space-time would incorporate the expansion rate of space-time indicating that as gravity is an effect it is still space-time traveling at over "c".

Regarding space-time curvature the concept of asymptotic flatness was proven when the ridiculously low figure of space-time curvature of 5 x10^-29 meters for a 10^54 kg universe making it appear as though space-time is flat(no curvature)except at the event horizon of black holes which have extreme density and gravitational effect. In this milieu the equation of everything was modified with the spiral operator $Ðn = 1to\infty \left(\dfrac{1}{2^{n\pi}} \right)$ *as space-time reduced from a sphere to an infinite curvature point is asymptotic to a spiral or a cone.*

So in terms of black holes $\mathbb{R}n = Ð\left(\dfrac{1}{2^{n\pi}} \right) g\ ab(R\ a\ b\ c - \dfrac{1}{2R} g \dfrac{ab}{\rho} ab$ *where* $\rho\ a\ b$ *was the energy density of matter from Poisson's Equation for the* Metric g ab incorporating the mc^2 from the Ricci tensor expression R a b into ρ ab. *The limits of this infinite product again is n=1 to infinity. The mass of a string via the Regee slope and tension of a string were applied*

To M Theory with the tension on a 4-brane by the anti gravitational effects of Dark Energy as a relaxation on the brane with duality to the concept of reciprocal curvature on space-time caused by the anti-gravitational metric of Dark Energy.

Finally there was a mathematical amplification of the mechanism of the rotational moments at or near the infinite momentum limit in the quantum bubble transforming it with centripedal force and acceleration to two poles such that the sphere became a double torus with antimatter at one pole and matter at the opposite pole and then a hyperbola with the 0,0 point(fracture point for the Big Bang with a symmetrical throat)causing an extreme rotation which unwound in 10^{-43} seconds to a pushing expansion of 6.75×10^{34} erg during the first second after "The Big Bang" with a trajectory of π *radians.*

"Spooky Action at a Distance" and variability of reading when an observer is present or absent are still up to some conjecture. Finally, whether there was a "Big Bang" or Inflation will depend on whether or not gravitational perturbations in space-time can detect a center of mass for the universe to prove isotropy or anisotrophy.

Also, the likelihood of a "Big Crunch" is remote as the required mass to meet the threshold to pop or rip space-time at the speed of light boundary "c" is so high that it would require virtually the entire mass of the universe 10^{54} kg to rip or pop space-time to cause a Big Crunch. This is based on space constricting and time dilating to infinity at the speed of light. Of course at "c" which is a boundary space-time approaches non-existence without reaching it and a boundary for space-time is a site where it can tear or pop.

FOOTNOTES AND BIBLIOGRAPHY

1. Peat, F.David. Superstrings and the Search for the Theory of Everything.Yang Mills Forces p.114

2. Kaku, Michio.Strings, Conformal Fields and M theory.Ising Model p.176-78

3. and 23:Maxwell's Equations and Stress Energy. Wald, Robert. General Relativity.Wikipedia.en.wikipedia.org/wiki/ Maxwell%27s –equations 4.CPT THEOREM; Quantum Field Theory Kaku, Michio

4. Metric tensor(General Relativity)Wikipediaa and Spacetime. en.m.wikipedia.org.spiral Space-time Einstein 1912 Fractal Time.p.108-109 Braden, Gregg 2009 Library of Congress HAWKING RADIATION. Wikipedia

5. Peat, F.David. Superstirngs and the Search for the Theory of Everything.p.106-107.Calabi Yau Manifolds

6. Kaku, Michio. Quantum Field Theory.p.648 19:Renormalization Actions in Quantum Field Theory

7. Peat, F.David. Superstrings and the Search for the Theory of Everything.p.156-161

8. Kay, David C. Tensor Calculusp.129 Osculating Plane

9. Kaku, Michio. Strings, Conformal Fields and M theory.p.505

10. Peebles, P.J.E.Principles of Physical Cosmology.p.365 and 392

11. Chang, Alan.HAMILTON JACOBI EQUATIONS UNIVERSITY OF CHICAGO 2013.Zeno's paradox: The Math Forum at Drexel University

12. Tipler, Frank j.The Physics of Immortality

13. Godel, Kurt.Godel's Incompleteness Theorems en.m. wikipedia.org

14. Randall, Lisa. Warped Passages p.60

15. Green, Brian.The Elegant Universe.and14.Wick, Mitchell Albert. Megaphysics, A New Look at the Universe. Introduction. 15:Kay, David.C.Tensor Calculus.p.129.

16. ibid item#7p.237-242

17. ibid item 16p.18

18. Kaku, Michio.Strings, Conformal Fields and M Theory.p.464-468

19. Wikipedia.Electronen.wikipedia.org/wiki/Electron

20. SpookyActionataDistance.QuantumEntanglementWikipedia. Or en.wikipedia.org/wiki/Quantum entanglement

21. Hau, Len.Harvard Research circa 2003.

BIBLIOGRAPHY

Barrero, John D.The Anthropic Cosmological Principle.Oxford England.Oxford Press.1986

Brade, Gregg.fractal Time 2009 Library of Congress.

Greene, Brian.The Elegant Universe.NewYork.Vintage Books editor Random Press.1999

Hawking, Steven and Penrose, Roger.The Nature of Space and TimePrinceton, N>J>Princeton Science Library 1996

Kaku, Michio.Quantum Field Theory.A Modern Introduction.Oxford university Press.1993

Kaku, Michio.Strings, Conformal Fiels, and M theory 2nd edition. Springer Press.2000.

Kay, David C.Tensor Calulus Schaum's Outline Series. N.Y.McGraw Hill 1998.

Peat, F.David.Superstrings and the Search for the Theory of Everything.Chicago.Contemporary Books 1998

Peebles, P.J.E.Principles of Physical Cosmology.Princeton Series in Physics.Princeton University Press 1993

Wald, Robert m.General Relativity.Chicago, Illinose.University of Chicago Press 1984

Wikipedia: on lin encyclopedia.

Randall,. Lisa. Warped Passages HarperCollins Publishers.N.Y.2005

Tipler, Frank J. Physics and Immortality. Anchor Books division of Random House.1993

APPENDIX

BRANES;A STRING IS A one-BRANE WHICH COUPLES TO A BACKGROUND SECOND DEGREE TENSOR.Zero-branes are ten dimensional building blocks for space in the pre-Big Bang epoch. The second degree tensor is purported of negligible mass as indicated by the ZERO-BRANE.THE SOURCE OF THE BACKGROUND SECOND DEGREE TENSOR IS R uv where the integral of D to the d power of x where x is the string or one-brane in D dimensions applies to R u v g u v where g u v is the metric acting on R u v for the zero-brane with respect to x which is the one-brane. R u v is the second degree tensor upon which the metric g u v acts. In four dimensions a monopole is dual to two electrons acting on a zero-brane. In 10 dimensions a string is analogous to a five-brane based on p-brane potentials. This involves dual fields such as a tensor R=R* from Ra1...a n=R8b1...b n.p-branes are encircled by a hypersphere which relates to M theory being compactified (curled up)by a circle for typeIIa strings.The charge of a p-brane is based on Q= $\int$*R *from limit S d-p-2 for electric charges and* $Q\int_{S_{p+2}} R$ *for electromagnetism. P-branes tie in with the potential involved with permutations of a field tensor.p brane tensors are associated*

Are associated with a tensor of the pth rank R a1...a p and electric and magnetic charges can be associated with p-branes with superalgebra.

Dzero-branes represent the vacuum state.ALTHOUGH INDICATED AS ten dimensional building blocks of space they actually are zero-dimensional. ONE-BRANES REPRESENT STRINGS WHICH ARE TWO DIMENSIONAL OR POSSIBLY ONE DIMENSIONAL. If all the dimensions in a system or universe are conserved such that the total number of dimensions are constant;then zero branes would have to be ten dimensional in the vacuum state.Six dimensions for CalabiYau Manifolds and four dimensions of space-time. As the "c" boundary is approached infinite mass with reducing length and width occur when length becoming infinite.In this case width and height approach zero but do not reach it and become infinitely small curled up and compactiifed.In general although showing duality between different systems which are abelian membranes are described by the forces involved with mass or energy associated with the membrane with reference of n-dimensional space where n dimensions would have n-1membranes or n-1 brane.

FLAT OR MINTKOWSKI SPACE IS DESCRIBED MATHEMATICALLY AS THE LINE ELEMENT OR ds2=dx2+dy2+dz2-c2dt2.FLAT SPACE-TIME IS SPACE-TIME WITHOUT ANY CURVATURE IN OTHER WORDS A VACUUM STATE HERE R g a b=0 which indicates that the space-time curvature metric=0 and therefore gravity =0 in the vacuum state.

CURVED SPACE-TIME IS GENERALLY DESCRIBED BY ds2=e-k|r|(dx2+dy2+dz2-c2dt2)+dr2 where r=space-time curvature metric described by tensor as R g a b.R g a b or r is determined by the inertial mass of the object doing the curving and the curving is performed by bosons and possibly gravitons or fermions.spiral space-time has k=-i(the square root of-1)to the n cotangent theta power as suggested by Dr.Roger Penrose and proposed by this author.

MANIFOLDS

THE SIMPLEST MANIFOLDS ARE CARTESIAN SPACES WHERE A MANIFOLD STRUCTURE OR SURFACE IN TERMS OF TOPOLOGIES IS R to the d power with what's called an identity map Rd implies R d. The coordinate functions of this map are cartesian coordinates.If coordinates are a I ;R d is the manifold of the standard Cartesian coordinates.a i=ax+ay+az and R to the d power is the tolological expression of the standard manifold or the Cartesian Coordinate system. If a manifold is imbedded in another manifold it is a submanifold.On a string basis submanifolds can be orbifolds or Calabi Yau manifolds which are submanifolds for spiral manifolds for asymptotically flat but curved space-time on a macrostatic surface which is expanding and simultaneous rotating as at black hole event horizon.

RIEMANNIAN CURVATURE A RIEMANNIAN SPACE IS THE SPACE COORDINIZED BY xi(power)with a fundamental form of the Riemannian Metric g I jdx I dx j where g=(g ij) obeys the metric tensor.g is of differentiability class C2(all second order partial derivatives of g I j exist and are continuous.g is symmetric g I j=g ji;g is nonsingular |g I j| doesn't equal 0.The differential form and distance from g isn't variant with regard to changes in coordinates.

R I j k l=g I ir(Rr superscript with jkl as a subscriptwhere R jkl with ias a superscript is the Riemann tensor of the second kind.The Riemann Tensor of the first kind is R I j k l=$\frac{\tilde{A}jki}{xk}-\frac{\partial \tilde{A}jki}{xi}+$ ΓilrΓjk with r as a superscript+ΓilrΓjk with r as a superscript-ΓikrΓji with r as a superscript. Here Γ *is a Christoffel symbol or the derivatire of a tensor. Above is* $Rijkl-\frac{\partial \tilde{A}jli}{\partial xk}-\frac{\partial \tilde{A}jkl}{dxl}+\tilde{A}ilr\tilde{A}jk$ with r as superscript-ΓikrΓjl withr as superscript.Skew Symmetrys involve Bianchi^' sIdentity R ijkl+Riklj+Riljk=0 skew symmetry is R i j k l=-Ri j kl an d second skew symmtery is R ijkl=-R i j l k with R j k l with i as superscript=-R j l k withi as a superscript. Block symmetry is R i j k l=R k l i j. These symmetry properties must fir with the $n2(n2-\frac{10}{12}$ COMPONENTS OF THE RIEMANN TENSOR(R i j k l)where the diagnal tensor without s.PTR.

Rijkl=g I iRjkl is subscript and I as superscript in the diagonal metric te tensor calculations for the Riemann Metricgives six cases R one R 212 and 1 R 313 and 1 R 323 and 1 R 213 and 1 R 232 and 1 R 123 and 1 which proves with the partial derivatives of the Christoffel symbols of tensors according to the previous formulas give R I j k l=0 for all I j k and l indicatin the summation of all Riemann forces and space is zero.The math of all these combinations is very difficult to reproduce by typing.

MATHEMATICAL FORMULAE

S=r(theta) the equation of arc length when applied to the osculating plane will pptro an infinite number of dimensions. The unit tangent vector of a sphere of 2

EXPANSION FACTOR. As

Down to theta approaches zero causes r or the arc length at the circumference to approach zero

CONTINUATION OF MATHEMATICAL EXPOSITION

kl

R ijkl - R ji=-8π G e (*ij,ji*) =g

-8 $\pi Ge(ij,ji)$ =g ji

vector product of Rij 'Rji=e (*ij,ji*) cos π =-e (*ij,ji*)

8 πG=Λ-1 is from cos θ θ=π klR ij.R ji ji for space-time curvature from antiparticle of metric g ji on antiparticles Rji kl=contra-variant tensor kl on covariant tensor ji ji is from antiparticle g ji kl is from gravity of contra-variant tensor for metric g ij for matter

Ijkl=antigravity effect on particles from antiparticles

2

R ij Rji   2   2

--. e ji where Rij

$$Rji/=\left\|\begin{matrix}R & ij\end{matrix}\right\|\left\|Rji\ \ \ \right\| = \cos\theta$$

$$\left\|\begin{matrix}2\\Rij\end{matrix}\right\|\ \ \ \left\|Rij\ \ \ \right\|$$

add e kl to e ij e kl= R kl / $\left\|\begin{matrix} R\ kl^2 & R\ kl \end{matrix}\right\|$ 'g kl cos 'g kl=-1. g kl where

$\theta=\pi$ *radians as covariant tensor For contravariant tensor e kl=R kl*

squared/ $\left\|\begin{matrix} R\ kl & R\ kl\ .\ g-g\ kl=g\ lk \end{matrix}\right\|$

kl

R ijkl = g kl/g ij'gkl=-8$\dfrac{\pi G}{-16\pi G}=\dfrac{1}{8\pi G}$ whioch is the reciprocal of

the cosmologic constant $\Lambda 8\pi G = 6.67x10\dfrac{11}{8\pi}=7.5x10$ 10 *joules-seconds*

which is a huge antigravity effect of Dark Energy from the antimatter

antimatter explosion of the Big Bang Q.E.D.

7

The gravitational coupling constant k=-8πG/c4 -8πG/c4 is also synonymous with antigravity between 2 particles of matter or anti-matter. As matter- matter interactions have a positive gravity anti-matter anti-matter interactions are antigravity Q.E,D.

Using E=mc2 with the number of strings in the quantum bubble for antimatter and matter being the mass the calculation emerges as E=7.5x 10 10 joules(3x10 10 cm/sec)2=6.75x10 31 joule-sec or on one second 6.75 x10 31 joules as the blast force from "The Big Bang" ;as antigravity is the predominant blast force Dark Energy blasted out in the 360 degree Orb Blast producing 750 billion galaxies from the Mas string calculation. Note 10 31 is 10 to the thirty-first power

4

NOTE: Although some sources list the cosmologic constant (Λ) as 8G/c this has no impact on the mathematical calculations only labeling. Other sources list this as the gravitational coupling constant.

G is in kg-meters/sec 2

6.67x10 11/8 =7.5 x 10 10 for number of strings. Calculation or at 10 7 erg per joule for 'Big Bang' blast force is e=mc 2 or e(joules)=7.5x10 10(3 x 10 10)2=75 x 10 30 joules or at 10 7 erg per joule 75 x 10 37 erg or 7.5 x 10 38 erg

mass are sequestered proximal to the event horizon so the extreme progressive curvature of spacetime to the infinite curvature of a string sized point is preserved.Therefore the net information going in and out of the black hole event horizon is a symmetric biconar surface with region 3 canceling region2 leaving regions 1 and 4 to produce the quasar effect.

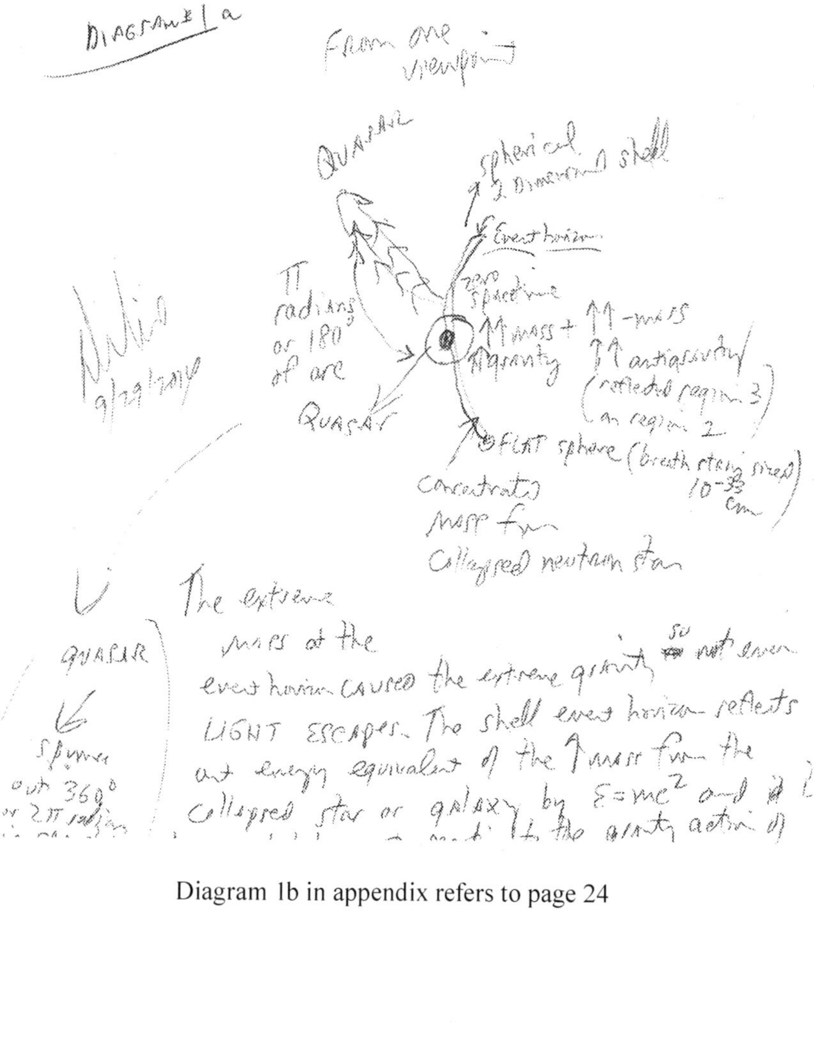

DIAGRAM 1 a

FROM one viewpoint

QUASAR

spherical $\frac{2}{2}$ Dimensional shell

Event horizon

zero spacetime

M mass + M - mass gravity $\overline{M}$ antigravity (reflected region 3) in region 2

FLAT sphere (breath string sized) 10^{-33} cm

Concentrated Mass frm Collapsed neutron star

π radians or $180°$ of arc

QUASAR

QUASAR

galaxy

The extreme mass at the event horizon caused the extreme gravity so not even LIGHT escapes. The shell event horizon reflects out energy equivalent of the $\uparrow$ mass frm the collapsed star or galaxy by $E=mc^2$ and $\#$ the gravity action of

QUASAR

sphere out $360°$ or 2π radians

Diagram 1b in appendix refers to page 24

Diagram #2

Quadric Surfaces in

vectr fm

$(Ar) \cdot v +$
$2 a \cdot v +$
$a_{44} = 0$

(1 D)   (1)   Quvat
                  Bubble
                  pre-Planck
                  Tme

(2 D)   (2)   (double tums)
                  ō stamp

(3 D)   (3)   (hyperbola)

FEYNMANN
Diagram   (0,0 point)

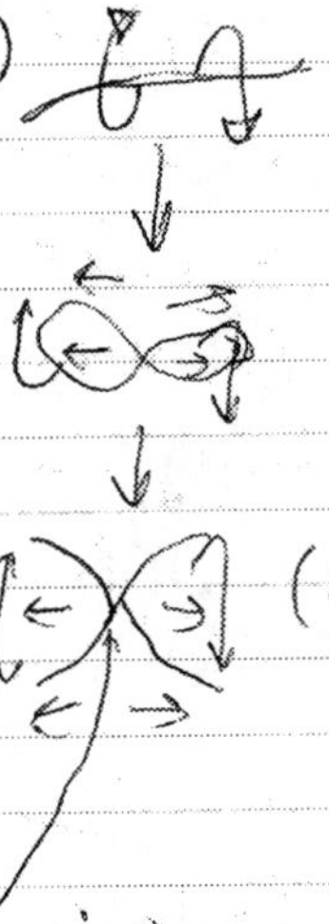

Feynman
Diagrams
for pre-BIG
BANG
Quantm
Bubble

Antmatter
matter mix

query would Antmatter move to
one extreme of hyperbola ō matter   DARK
  at other extreme   BLAST Energy
                                matter (ANTI matter) matter DARK energy
  or
Antmatter   will they stay mixed
+ matter   IN A homogeneon mixture   BLAST
  matter (50) matter
  Antimatter ANTIMATTER

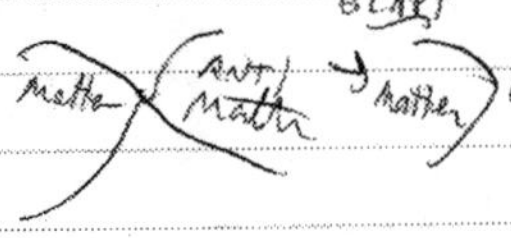

Antimatter
+ matter
ō
Antmatter
+ matter

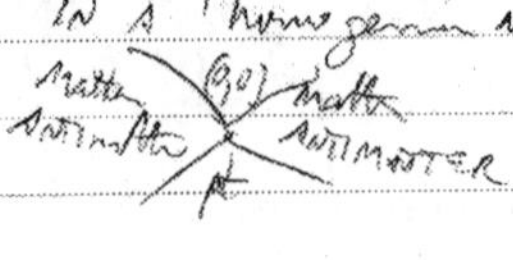

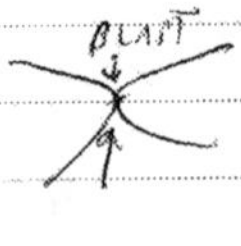

Diagram
#3

Feynman

Diagram at
0,0 pt

Matter   ANTImatter

Symetts THROAT
Standerd
              matter (Anti)
        matter   matter

DARK Energy
ANTIGRAVITY

energy
from
Matter —
ANTImatter
ANNIHILATION

Orb blast
related to isotropic
homogen universe

Dark matter
residual from
from ANTImatter

oromos
Matter
left
to form
Matter
Universe

- from BLAST
0 Antimatter —
matter
ANNIHILATION
purity of ANTI
particles
Away from each
other given
out as DARK
Energy

4/9/2025

Diagrams 2

Quantum Surfaces in

vector form

$4(r) \cdot r +$
$2 a \cdot r +$
$a_{44} = 0$

(1D) ① — First Bubble, pre-Planck Tube

(2D) ② — (double time) ε strings

(3D) ③ — (hyperbole)

Feynman Diagram

(0,0 point)

Feynman Diagram for pre-Big Bang Quantum bubble

antimatter / matter mix

query would antimatter move to one extreme of hyperbole ε matter to other extreme
or
will they stay mixed in a homogeneous mixture

matter / antimatter
matter / antimatter

over energy
over energy

Averaging
ε matter
ε
antimatter
ε matter

USING SCHWARZCHILD SPACETIME AT BLACK HOLE EVENT HORIZONS TO DISPROVE 'THE INFORMATION PARADOX

BY DR.MITCHELL ALBERT WICK

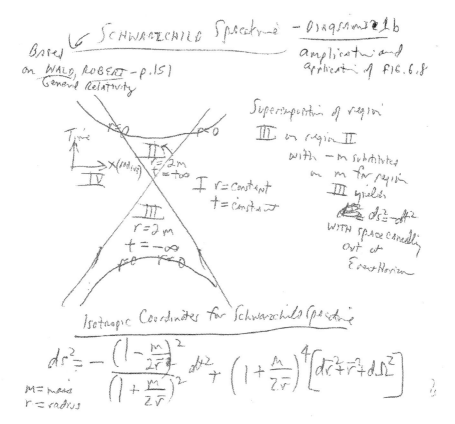

Based on WALD, ROBERT – p.151 General Relativity

SCHWARZCHILD Spacetime – Diagram #1b

amplication and application of FIG.6.8

Superimposition of region III on region II with −m substituted on m for region III yields

$$ds^2 = -dt^2$$

WITH space cancelling out at Event Horizon

Time

X(radius)

IV

I r=constant t=constant

II r=2m t=+∞

III r=2m t=−∞

Isotropic Coordinates for Schwarzchild Spacetime

$$ds^2 = -\frac{\left(1 - \frac{m}{2\bar{r}}\right)^2}{\left(1 + \frac{m}{2\bar{r}}\right)^2} dt^2 + \left(1 + \frac{m}{2\bar{r}}\right)^4 \left[d\bar{r}^2 + \bar{r}^2 d\Omega^2\right]$$

m = mass
r = radius

GLOSSARY

Abelian:equations having a coefficient or variety in a specific group, g,,algebraic number fields, tensors of the same degree or cohominy group

Anisotropic:not isotropic, lacking observational symmetyry

Anti-symmetric:tensors or vectors that are equal but opposite and can therefore partially cancel or cancel

Aymptotic:that which approaches a level or degree but never reaches it;asymptotic flatness appears without curvature but doesn't reach it

Bianchi's Identity:The identity of groups of Riemannian 4 space that is anti-symmetric and Abelian and cancels each other out of being equal but opposite

"The Big Bang" A theory proposed describing a Friendman type I open expanding f;at universe with is homogeneous and isotropic

"The Big Swirl" A Big Bang with a progressively decreasing rotational vector from an infinite curvature point of space-time to asymptotic flattness

Black Hole:collapsed matter from a neutron star or galaxy with extreme curvature of space-time at the central nexus due to extreme gravity of of the spiral space-time

Calabi Yau Manifold:a surface which represents a relative isotropic portion of space-time with a puckering to accommodate multiple dimensions considered a twisted variant of the orbifold

Choas:absolute disorder

Chiral:a mirror image or absolute symmetry

Closed string:a two or one dimensional building block of matter from energy with movements in 10 or 26 dimensions without breaking the string

Compactified:when every point of the dimensions are curled up mathematically making the size approach zero.First determined by Kaluza and Klein

Conformal Space:when every point in space relative to every other point maintains its relative position regardless of what the space is doing

Dark Matter;an indirectly measured mass causing perturbations in gravity(the curvature of space-time)caused by mass.Acts as cosmic glue containing possibly baryonic particles and neutrinos

Glossary page 2

Event Horizon:area where a black hole is perceived by measurementsEntropy:degree of disorder

Entropy:degree of disorder

Ex nihilo:out of nothing

M(Membrane)theory:the 5 dual string theories into one massive theory of everything which incorporates membranes which vibrate and incorporate all energy and matter

Isoropic:observational symmetry

Geodesic:a unit of space-time

Gravity:the curvature of space-time caused by mass;actually an effect not a force

Membranes:a description of matter in terms of energy states with stress energy densities described in the number of states with regard to dimensions

N;number of dimensions in N dimensional space

Open string:a two or one dimensional bulding block of matter with movements in a multidimensional plane

Orbifold:space-time manifold in an open twisted cone configuration utilized in string theory

Relativity:the behavior of matter and energy with regard to other matter and energy;energy and space have a different vantage point from other matter and energy including stress energy, time and mass with changes regarding relative velocity

RicciTensor:that tensor which represents inertial mass or resistance against pull or push

Riemann Forces:all strong and weak forces in nature

Riemannian Space:Mintowski space with Riemann curvature of space-time caused by mass.Flat space if no mass is present

Scalar:the magnitude compone t of a vector or tensor with regard to direction

Space-time Manifold:Matrix over a Riemann surface in which the energy density of matter exists as per Poisson's Equatio, The derivative operator of a dual vector field is 4(pi)rho where rho is the energy density of matter

Stess:energy density with components of the Strong Force(quarks in the nucleus of an atom)and weak forces including radioactive decay, electromagnetic force(strong), inertia.

Tensor:a force vector with subcomponenets interacting with or without other vectors in either a moving or stationary frame where the magnitude only is called a scalar trace

Weyl'sTensor:a pull or comglomerate pull in a conformal matter>The pull represents space-time curvature which is the effect of gravity caused by mass

X or r:any observable in space-time

YangMills Notation:diagrammatically configuring the different forces of nature and the component masses which may or may not interact

Impossible:That which is beyond the comprehension of mankind

Science:a process to obtain knowledge

Printed in the United States
By Bookmasters